KU-498-910

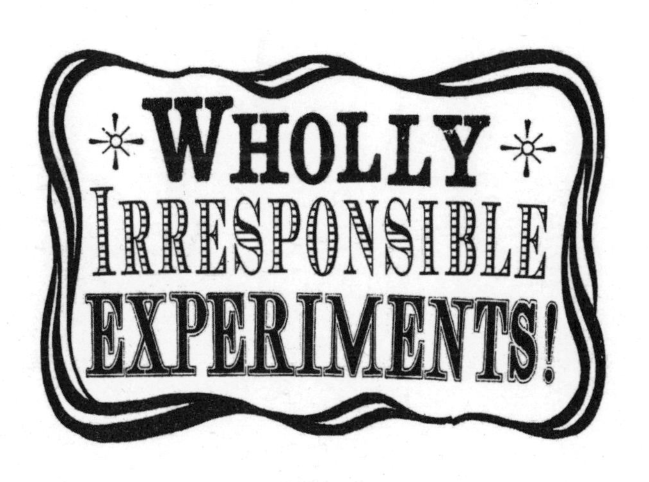

ICON BOOKS

Published in the UK in 2007 by
Icon Books Ltd, The Old Dairy,
Brook Road, Thriplow,
Cambridge SG8 7RG
email: info@iconbooks.co.uk
www.iconbooks.co.uk

Sold in the UK, Europe, South Africa and Asia
by Faber & Faber Ltd, 3 Queen Square,
London WC1N 3AU
or their agents

Distributed in the UK, Europe, South Africa and Asia
by TBS Ltd, TBS Distribution Centre, Colchester Road
Frating Green, Colchester CO7 7DW

This edition published in Australia in 2007
by Allen & Unwin Pty Ltd,
PO Box 8500, 83 Alexander Street,
Crows Nest, NSW 2065

Distributed in Canada by
Penguin Books Canada,
90 Eglinton Avenue East, Suite 700,
Toronto, Ontario M4P 2YE

ISBN-10: 1-84046-812-2
ISBN-13: 978-1840468-12-0

Text copyright Sean Connolly © 2007

The author has asserted his moral rights.

No part of this book may be reproduced in any form, or by any means, without prior permission in writing from the publisher.

Typesetting and design by Simmons Pugh

Printed and bound in the UK by Clays

ADULT SUPERVISION NEEDED

THESE EXPERIMENTS INVOLVE MATCHES, HOT LIQUIDS AND INGREDIENTS THAT COULD BE HARMFUL IF USED INCORRECTLY

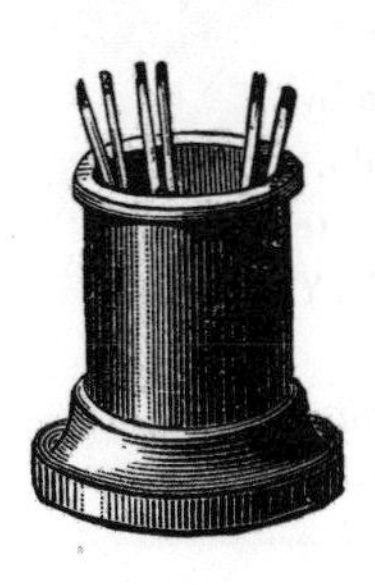

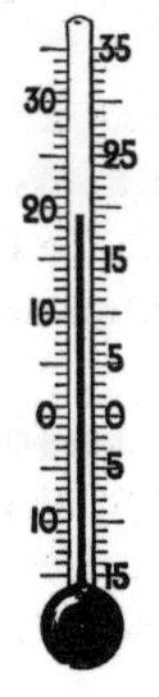

T
B
R

To my companions on this wonderful journey –
Frederika, Dafydd, Jamie, Anna and Thomas

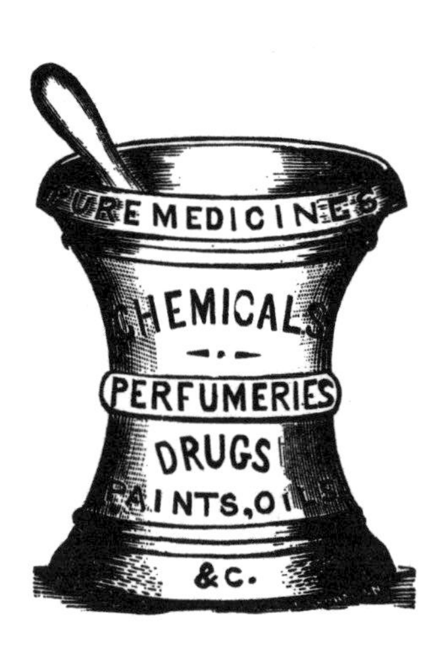
PURE MEDICINES
CHEMICALS
PERFUMERIES
DRUGS
PAINTS, OILS
&C.

Contents

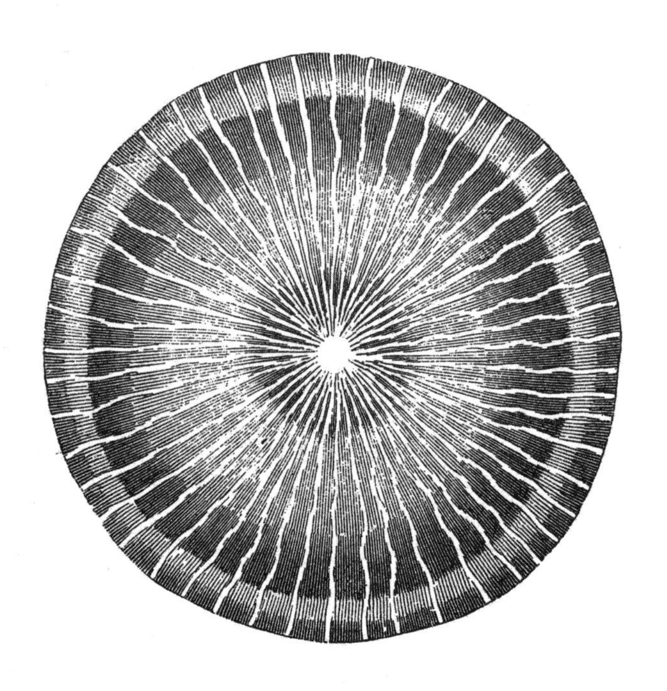

INTRODUCTION

The *Oxford Dictionary of English* defines science as 'the intellectual and practical activity encompassing the systematic study of the structure and behaviour of the physical and natural world through observation and experiment'.

This definition explains the link between mankind's earliest paintings – with their vivid depictions of cave lions and predatory wolves – and the NASA space shuttle studies of atmospheric winds using laser radar. Down through the ages, people (whether or not they termed themselves 'scientists') have observed and experimented their way to a better understanding of the world and how it works.

Throughout our history, we have been driven by curiosity and the 'need to know'. Scientists have probed all manner of conundrums, teasing out answers and sharing their findings. Just think how much of our knowledge can be traced to questions such as:

Why does the sun rise and set every day?

Why did that ripe apple fall down from the tree and not up?

If water expands when it becomes steam, could it be used to drive a piston?

Could more than one computer be linked together using, say, a telephone connection?

We know – or know how to find out – the answers to these and thousands of other questions that have inspired scientists for centuries. And we can see the benefits all around us, especially in the field of technology, which harnesses the advances of science and turns them to practical advantage.

NEW SETS OF QUESTIONS

Wholly Irresponsible Experiments carries on this noble tradition of scientific exploration and takes it to new – yet, in many ways, familiar – areas. After all, the quest for knowledge does not end when we hang up our goggles and turn off the light in the science lab. The everyday world provides us with the tools to carry on with our scientific probing.

The 65 experiments described in the following pages use ingredients or materials that most households have, or which can be bought easily. And like the classic scientific experiments, which use questions as launch-pads for enquiry, these experiments also seek to find and demonstrate answers. Some of the answers, however, might well tie in with a completely different set of questions. Along the lines of:

'What's that straw doing inside a potato?'

'Darling, have you seen my hairdryer?'

'What in the world has happened to this geranium?'

'Wait a minute! Is that my new Coldplay CD floating around over there?'

The 'i' word

All of this brings us to an important word in the title of this book – 'Irresponsible'. Where does being irresponsible tie in with conducting experiments? Surely it's the opposite of the scientific method? In response we can cite former US President Bill Clinton and his famous phrase: 'That depends upon what your definition of "is" is.' Well, if 'is' can carry with it more than one meaning, then it's hardly surprising that there could be more than one reading of the word 'irresponsible'.

Although using the 'i' word, *Wholly Irresponsible Experiments* advocates due care and attention in each experiment. The presentation of each experiment is straightforward and logical, right down to any words of special warning (in the Take care! section) that apply to the experiment. Instead it calls on readers to throw off some of the shackles of being a grown-up and find the child within us all.

No child, after all, would worry about a sharp smell filling the kitchen or black smoke billowing from the back garden. And that inner child would hardly prefer sorting out an income tax return (or any other 'responsible' activity) to letting loose a two-stage balloon rocket. So, by that definition, each of the experiments certainly does merit the descriptive term 'irresponsible'.

Who can do these experiments?

Any book that appeals to the 'child within us' is bound to appeal to children themselves, and this is no exception. We heartily recommend demonstrating each experiment to budding young scientists, and some of the experiments can involve children as volunteers or even participants. But bear in mind that the responsibility for each experiment lies with the adult conducting it. These experiments are *for* children as well as adults, but they are not to be conducted *by* them.

The final section of each experiment, Take care!, highlights any particular warnings relevant to the experiment. Some of these are no more than bits of friendly advice on how to get the best effects. Others have a more practical aim, of drawing the reader's attention to ingredients or actions that call for extra care. A special Match alert is a prominent flag to any experiment that involves matches or an open flame.

Some people might feel that they've left it too late to return to science. Maybe they were bored back in science class at school, or perhaps they felt that it was all a bit too difficult to grasp. *Wholly Irresponsible Experiments* offers them a chance to re-enter this fascinating world. Apart from producing a result that will amuse, enchant or possibly even inspire, each experiment is presented in a form that most of us recognise: a simple recipe.

How this book works

The 65 entries in *Wholly Irresponsible Experiments* are grouped in seven chapters, each representing a different scientific theme or intended result.

A typical entry introduces the nature of the experiment and what to expect, before breaking it down into the following sections.

YOU WILL NEED – A straightforward list of ingredients.

METHOD – Numbered step-by-step and easy-to-follow instructions.

THE SCIENTIFIC EXCUSE – The *raison d'être* for the experiment – or possibly your hurried explanation to an impatient or angry spouse!

TAKE CARE! – Special advice (and in some cases, warnings) for the experiment.

AT A GLANCE

The 65 experiments have also been grouped at the back on p. 168 according to how long it takes to complete them – from the first stage of preparation to the 'oohs' and 'aahs' at the conclusion. You might have a whole Saturday at your disposal or only a few minutes free.

The categories here help you to choose an experiment perfect for the time you have spare.

FLASH IN THE PAN – less than 2 minutes

FIVE-MINUTE WONDERS – 2–5 minutes

ON THE HOUR – up to 1 hour

THE 8-HOUR DAY – 1–8 hours

GOING THE DISTANCE – a full day or more

Isn't it time you went out and built that volcano you've been promising yourself? Or maybe you're worried that your marshmallows aren't macho enough? Perhaps you've decided you're ready to change the colour scheme in the sitting room, *starting with the plants that are already there*! The following pages will let you do all of these things, and much more, all in a spirit of playful scientific enquiry.

For most of the experiments, a broad smile and an open mind will count for far more than a white coat and a calculator. So throw yourself into these funny, eye-opening, quirky experiments and see where they take you. And in the process you'll have a chance to teach – or maybe learn – a little science!

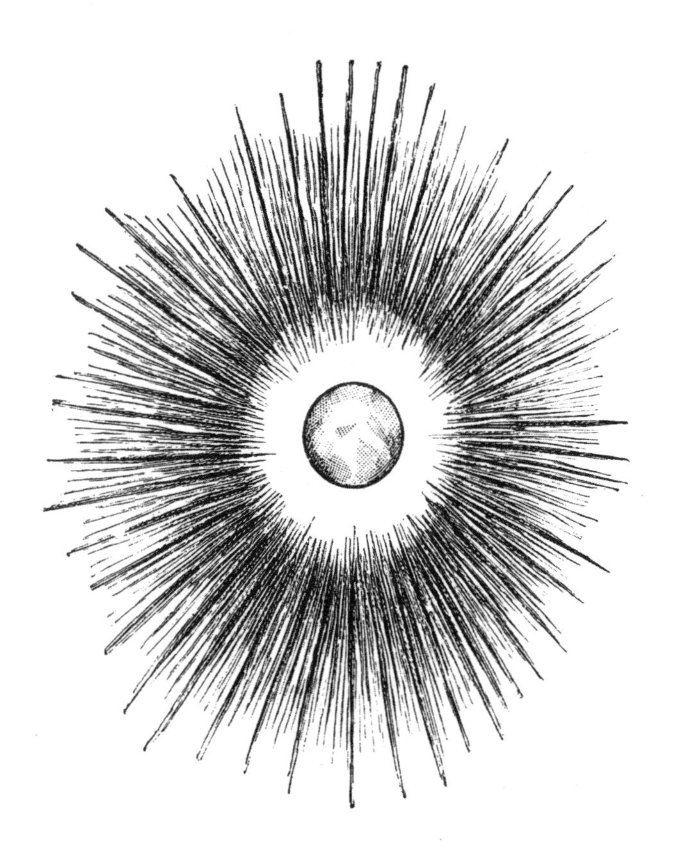

Core Concerns

At times the Earth seems like a child trying to wriggle from a coat that's just too small. Chasms open, the ground shakes with quakes and tremors, while magma, lava, hot water and steam burst through the planet's crust with dramatic effect. The experiments in this chapter offer a glimpse at the basic physics that lies behind these 'earth-shattering' events. Cooking ingredients, basic kitchen equipment and even sweets can all play their part in explaining some of these basic forces.

The Cola Geyser

Some of the most memorable experiments can be done using ingredients that don't seem in the least 'scientific'. For example, you can mix a popular sweet and an even more popular soft drink to create your own version of Old Faithful. The volatile mixture sends a geyser as high as 15m.

METHOD

1. Put 12 Mentos in a test tube and hold the card to the open top of the tube.

2. Open the bottle of diet cola and put the test tube upside down on the top of the open bottle, still holding the card in place.

3. Take care to know which way to run.

4. Slide the card away quickly so that the Mentos drop in.

5. Run clear and watch as the cola explodes out of the bottle.

THE SCIENTIFIC EXCUSE

This explosive reaction comes from the sudden release of carbon dioxide, the gas that gives fizzy drinks their fizz. This carbon dioxide normally remains dissolved in the soft drink because there are no nucleation sites – irregularities around which bubbles can form. A single Mento, seen close up, has a craggy surface – providing hundreds of nucleation sites. A dozen of those sweets, dumped in at once, sets off a massive release of carbon dioxide, forcing the cola out of the bottle like a rocket. Diet cola works best because most non-diet colas use corn syrup, which suppresses the formation of bubbles.

TAKE CARE!

Do this experiment outside, well away from anything (or anyone) that you wouldn't want to become a sticky mess.

Back-garden Vesuvius

This vivid display of a chemical reaction isn't dangerous, but it earns its place in this book by being very, very messy. It should go without saying that this is an outdoor experiment, so make sure you choose a dry day to demonstrate those dramatic lava flows. And if you're really feeling resourceful, you can add to-scale model houses at the base of the volcano. Make sure you don't clap when they get swept away.

· YOU WILL NEED ·

- 75 cm x 75 cm plywood sheet (or larger)
- Empty 1-litre bottle (glass or plastic)
- Plasticine or papier-mâché (enough to form a 'volcanic cone' about 40 cm across at its base)
- 1 tbsp bicarbonate of soda
- 15 ml washing-up liquid
- Red or yellow food colouring
- 60 ml vinegar

METHOD

1. Put the empty bottle in the centre of the plywood sheet. It will be the centre of the volcano.

2. Build the volcano around this bottle, using either Plasticine or papier-mâché.

3. Work on and decorate the volcano while it's still soft, carving out gulleys and ravines for the lava flow.

4. Make sure the finished volcano has enough time to dry.

5. Add the bicarbonate of soda, the washing-up liquid and a squirt of food colouring to the empty bottle.

6. Measure out the vinegar and pour it into the bottle.

7. Stand back as the forces of nature take over!

THE SCIENTIFIC EXCUSE

The trigger for the eruption is the addition of the vinegar (an acid) to the earlier mixture, which is basic thanks to the inclusion of bicarbonate of soda. Adding the vinegar triggers an 'acid/base neutralisation', as chemists would put it. More specifically, this reaction changes carbonic acid into water and carbon dioxide; the liberated carbon dioxide leads to the dramatic foaming.

TAKE CARE!

The biggest problem with this experiment is the possible mess, but giving too much of a warning here would be overdoing it, wouldn't it?

The Sandwich-Bag Bomb

Kids – and many grown-ups – may think of acids and bases as harmful chemicals that white-coated scientists keep locked away. This experiment is a simple way of showing that a kitchen cupboard has these chemical 'secret agents' in abundance. And with just a little outlay in time and equipment, you can get 'more bang for your buck', as the Americans say.

METHOD

1. This experiment will work only if your sandwich bag has no holes. Test it by half-filling it with water, zipping it shut and turning it upside down over the sink.

2. If no water leaks out, you're fine. Empty the test water out.

3. Tear a sheet of paper towel into a square measuring about 15cm x 15cm.

4. Pour the bicarbonate of soda onto the centre of the paper towel, then fold the towel over itself with the powder inside it.

5. Pour into the bag the vinegar and warm water.

6. Then, carefully but quickly, add the paper-towel 'envelope' to the bag and seal it.

7. Shake the bag a little and then put it on the ground and stand back.

8. The bag will inflate and then pop with a satisfying bang.

THE SCIENTIFIC EXCUSE

The vinegar and bicarbonate of soda react dramatically and quickly, producing carbon dioxide equally quickly as a result. This carbon dioxide soon fills the bag and then, after straining at the bag's seams, pops it with a bang.

TAKE CARE!

No one is going to get hurt in this experiment but it can produce a mess. Choose your spot carefully beforehand, making sure nothing – or nobody – will get soaked or soiled!

Flameproof Balloon

Here is another of those 'less is more' experiments. With precious little investment (how much does a balloon cost, after all?) and no real preparation, you can perform an experiment that is a real eye-opener. Balloons are always popping, aren't they? Yet why won't this one burst when it's placed in an open flame?

• You Will Need •

- Balloon (strong enough to hold water)
- Candle
- Matches
- Water
- 2-litre drink bottle full of water (optional)

Method

1. Fill the balloon with water until it is about the size of a grapefruit. You can do this by fitting the mouth of the balloon over a cold-water tap. Alternatively (if you're away from the kitchen), you can do the same thing over the mouth of the drink bottle.

2. Tie the balloon with a secure knot and shake it so that any excess drops fall off.

3. Light the candle and place it somewhere secure.

4. Hold the balloon by pinching the knotted end and move it into the flame.

5. Keep the balloon in the flame for a few seconds, then remove it.

6. Repeat the process – the balloon seems flameproof!

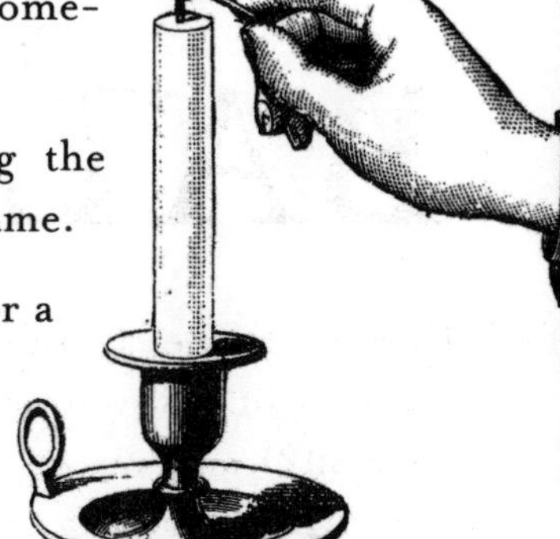

— THE SCIENTIFIC EXCUSE —

This experiment tells us a lot about heat absorption. If you held a 'normal' balloon (filled with air) in the flame it would burst quickly. The heat of the flame would weaken the rubber until it could no longer withstand the pressure of the air inside. Water, however, absorbs heat very well. This means that the water inside the balloon absorbs the heat of the flame, leaving the rubber more or less unscathed.

TAKE CARE!

Blow out the candle after performing the experiment.

MATCH ALERT!

This experiment involves the use of matches and should be conducted only by a responsible adult.

Beacon from Atlantis

The advanced civilisation of Atlantis vanished beneath the waves before the time of the ancient Greeks. With it went – so the story goes – a wealth of knowledge, never to be recovered. But maybe not all of these skills have disappeared. How else could you explain a candle staying alight, even when it's under water? It must be a message from below.

· YOU WILL NEED ·

- Small mixing bowl (preferably clear)
- Candle (as tall as the bowl is high)
- Matches
- Water

METHOD

1. Heat the base of the candle and secure it inside the bowl in the centre.
2. Carefully fill the bowl with water right up to the rim of the candle.
3. Light the candle and observe.

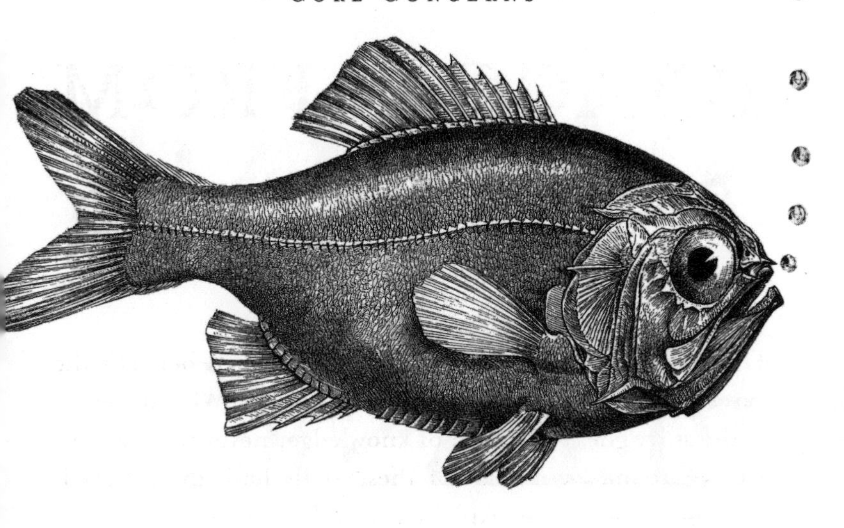

4. The candle wick burns down, even below the water level, and forms a wax funnel around itself.

5. The flame, protected by the funnel, continues to burn for some time below water level.

— THE SCIENTIFIC EXCUSE —

Normally, in the open air, the wax by the wick would melt and either evaporate or flow down the side of the candle. But water is excellent at absorbing heat, and it draws the warmth that would otherwise melt the wax. So instead of melting, the wax remains its solid funnel shape, guarding the flame like a dyke against the water outside.

MATCH ALERT!

This experiment involves the use of matches and should be conducted only by a responsible adult.

Can you hold, please?

There's a little bit of magic in every scientific experiment, whether it concentrates on the tiniest particles or the speed of light through the infinity of space. This experiment inhabits that land of wonder and mystery, beginning as a puzzle and usually ending up with applause and laughter. Maybe it's best if you frame it as a challenge: can you pick this ice cube up from the pie plate without getting your fingers wet?

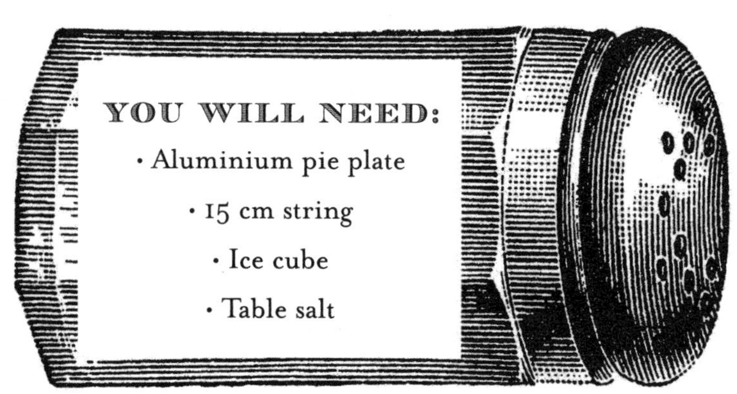

YOU WILL NEED:

- Aluminium pie plate
- 15 cm string
- Ice cube
- Table salt

METHOD

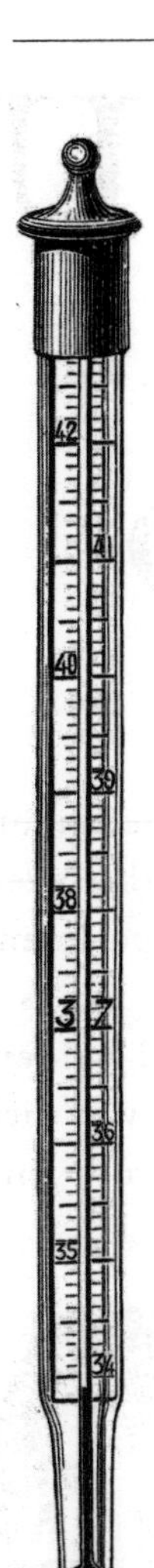

1. Rinse the ice cube and set it down on the foil pie plate.

2. Lay the string across the ice cube.

3. Sprinkle salt over the top of the ice cube, making sure much of it lands on or near the string.

4. Wait 15 seconds and then hold the two ends of the string.

5. Slowly lift the string, which will be holding the ice cube.

THE SCIENTIFIC EXCUSE

Salt lowers the freezing temperature of water (which is why roads are salted in winter). In this case, the salt causes some of the ice at the top of the cube to melt. The salty water is absorbed by the dry string, which lets the water at the top of the cube freeze over again. Except this time it has locked a length of string beneath the new layer of ice.

TAKE CARE!

Make sure you lift the two ends of string gently since a sudden jerk will almost certainly yank the string from the protective layer of ice.

Harnessing the Elements

When Freddie Mercury sang 'thunderbolts and lightning, very very frightening' in 'Bohemian Rhapsody', he hit a raw nerve. Since the dawn of history, human beings have been fascinated and frightened in equal measure by the power of the elements. And while we have a better explanation for these phenomena than, say, the ancient Greeks, we still have a hard time linking them to everyday forces around us. The following experiments call on familiar ingredients – flour, water, pie plates, biros – to demonstrate how weather works.

Home-made Lightning

What's weather talk without a little exaggeration now and then? OK, well maybe real lightning is a few trillion times more powerful, but this experiment goes right to the heart of the science that produces a lightning bolt – an electrical discharge. In the case of real lightning, this discharge goes from cloud to cloud. Here the distance and scale are more modest – across the breadth of a fingernail. But with the lights out, and a good explanation, this can be a real crowd-pleaser.

You Will Need

- 30 cm x 10 cm x 3 cm polystyrene block
- Biro
- Glue
- Drawing pin
- Aluminium pie plate
- Woollen sock

Method

1. Push the drawing pin from the back of the foil pie plate through the centre.

2. Press the non-writing end of the pen onto the tack,

securing with glue if needed.

3. Rub the polystyrene block quickly with the woollen sock.

4. Pick up the pie plate with the pen (not touching the pie plate itself).

5. Put the pie plate down carefully on the polystyrene.

6. Turn out the lights and draw your finger closer and closer to the pie plate.

7. You should see, hear and feel a small spark.

— THE SCIENTIFIC EXCUSE —

The rubbing causes (negatively charged) electrons to flow from the wool to the polystyrene, giving it a negative charge. Similar charges (positive and positive, or negative and negative) repel each other, so the electrons of the polystyrene cause some of the electrons in the pie plate to move away from the polystyrene. They are waiting to escape from the pie plate, but cannot move through the pen (because it is an insulator). They can, however, flow through the human body and jump across the small gap to reach the experimenter's finger.

TAKE CARE!

The following is not a safety warning (this is one of the safest experiments in the book). File it instead under 'follow this advice if you want the experiment to work'. Make sure you are holding the pen when you place the pie plate on the polystyrene. Otherwise, the excess electrons will flow undramatically from polystyrene to pie plate to finger.

Heavy Weather?

Global warming, carbon emissions, the ozone layer – it can all seem a little rarefied, especially for young people with relatively little science background. But give them a chance to create some carbon dioxide – and then to see how it's a force to be reckoned with – and they might start taking carbon offsetting a little more seriously. Plus, there's always something spooky about invisible forces at work around us. Build yourself a home-made scale to prove that the invisible carbon dioxide is heavier than the equally invisible air.

YOU WILL NEED

- 2 clear plastic bags (sandwich-bag size)
- Sticky tape
- 30 cm ruler
- 3 tsp bicarbonate of soda
- Drawing pin
- 100 ml vinegar
- Drinking glass

METHOD

1. Secure the drawing pin (pointing up) to a surface, near the corner of a counter or table.

2. Tape the plastic bags (leaving a good upward-facing opening) to each end of the ruler.

3. Carefully balance the ruler on the drawing pin to create a set of home-made scales.

4. Mix the vinegar and bicarbonate of soda in the drinking glass.

5. When this mixture begins to froth, tilt the glass carefully over one of the open plastic bags. Do not pour any of the liquid or even the froth.

6. The bag beneath the tilted glass should slowly sink under the weight of the invisible new ingredient.

— THE SCIENTIFIC EXCUSE —

A diligent science student will recognise that mixing vinegar and bicarbonate of soda is one of the best – and easiest – ways of producing carbon dioxide. Carbon dioxide is a by-product of the reaction between the acid (vinegar) and base (bicarbonate of soda). Moreover, carbon dioxide is heavier than air. That's why it can be poured, just as one could pour a liquid, into one of the bags, causing the scale to tip.

TAKE CARE!

Tilt the glass gently, once you have mixed the vinegar and the bicarbonate of soda. Having gone through the bother of creating the delicate scale, you don't want to blow the experiment by pouring the liquid into the bag – the whole point is that the bag is dragged down by the invisible gas!

STATIC ELECTRICITY SLIME

This is one of the best experiments for younger scientists, with the messy and seemingly risky combination of ooze and electricity proving irresistible. See how many 'bright sparks' can work out what's happening. To give them a clue, try creating Home-made lightning (p. 40) first.

· YOU WILL NEED ·

20 cm x 20 cm x 3 cm polystyrene block
500 ml vegetable oil
Large drinking glass (a pint glass is ideal)
100 g cornflour
Woollen sock
Refrigerator

METHOD

1. Mix the oil and the cornflour in the glass.

2. Put this mixture in the refrigerator until it's well chilled.

3. Remove from the refrigerator and stir (don't worry if it has separated).

4. Let this mixture warm enough so that it can flow; it will resemble a slime at this point.

5. Rub the block of polystyrene on the woollen sock (or on your hair if there's no sock to hand).

6. Tip the container of slime and put the electrically charged polystyrene about 2 cm away from the slime. It should seem to stop flowing and even to gel.

7. Try wiggling the polystyrene: bits of the slime might break off and follow it.

8. You can refrigerate the slime in a sealed container after the experiment.

— THE SCIENTIFIC EXCUSE —

When rubbed, the polystyrene draws electrons from the wool (or hair), giving the polystyrene a negative charge. Meanwhile the oil and cornflour combine to make a substance known as a colloid. When the charged polystyrene nears the colloid, it causes the cornflour to line up, blocking the flow of the liquid oil. Taking the polystyrene away lets the mixture behave more like a normal liquid again.

TAKE CARE!

This is one of those rare experiments that's safe enough to recommend with few reservations. Just remember to make sure that your glass is big enough to hold the oil–cornflour mixture.

Giant Air Cannon

Simple ingredients, easy to make, dramatic results – some experiments queue up to be included in this book. This air cannon is a lot of fun to make, especially with young children. You can crank up the 'irresponsible' meter dramatically with flour or other non-hazardous powders to add to the effect.

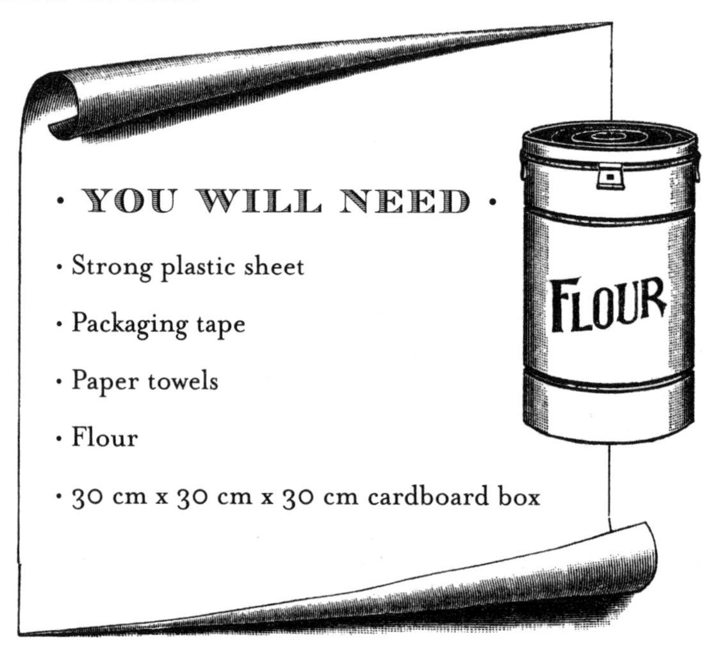

· You Will Need ·

- Strong plastic sheet
- Packaging tape
- Paper towels
- Flour
- 30 cm x 30 cm x 30 cm cardboard box

METHOD

1. Remove flaps from one end of the box. At the other end, tape together the flaps and cut out a single 15 cm circle.

2. Reinforce this circular opening with more tape.

3. Tape the plastic sheet over the open end, leaving plenty of slack; allow it to overlap each side by 12 cm before taping it securely.

4. Pinch the slack plastic sheet in the middle and then punch hard to operate the air cannon. You should get a satisfying whoosh of air.

5. Try putting some flour on the base of the cannon and then having a go. It should produce a blast of 'smoke'.

THE SCIENTIFIC EXCUSE

No matter how you doll up this experiment – with flour or with some other harmless powder – it's still easy to understand. With the plastic sheet kept slack, the cannon has a greater volume, and this naturally fills with air. The sudden punch reduces this volume quickly, causing the excess air to rush out.

TAKE CARE!

Not surprisingly, this is one of the least risky experiments in the book.

Burning ice

Lighting a fire with a block of ice? Surely that's a contradiction in terms! Try this experiment to see for yourself how the forces of nature can overturn common sense. It's certainly one display that you and your audience won't forget too easily.

YOU WILL NEED:

- Shallow, evenly curved plastic bowl (about 20 cm wide)
- Room enough on a freezer shelf for the bowl
- Black crepe paper
- Water

METHOD

1. Boil a saucepan of water for 3–4 minutes and let it cool to just above room temperature.

2. Fill the bowl with the previously boiled water.

3. Place the bowl of water on a shelf in the freezer and allow it to freeze completely.

4. Scrunch up a small piece of black crepe paper and set on a flameproof plate.

5. Remove the ice from the bowl (running a little warm water over the back of the bowl can help to release it).

6. Hold the ice, which is shaped like a lens, above the paper and direct the sunlight at the paper.

7. Keep the ice still until the paper begins to ignite.

THE SCIENTIFIC EXCUSE

This experiment operates under the same principle as letting a magnifying glass light an object: it concentrates, or refracts, the sunlight and focuses it on the paper. The ice has very little chance of cooling these rays. And boiling the water first is important as a way of 'purifying' the lens. Ordinary water has many bubbles inside it: even the tiniest ones (too small for us to see) can distort the lens and make it less effective.

TAKE CARE!

There are few risks tied in with this experiment. You needn't use a large piece of paper – the whole point is just to show how powerful the lens could be. Plus, there's no real fire risk if you let the paper catch fire, then make sure you put it out immediately.

FOOD FOR THOUGHT

It's amazing to think that just about every branch of science can be studied in the kitchen, using ingredients from the store cupboard, refrigerator, larder or cutlery drawer. Billowing black smoke, flying potato pieces, a mysterious growing hand – these things are the real stuff of science, aren't they? You can eat some of the results, but you might find that others have simply disappeared. All of this gives a wild new twist to the term 'domestic science'.

Bacon Smokescreen

Here's a way to indulge your wish to enter the world of smoke and mirrors. Smokescreens are popular with young and old, and the modicum of science that underpins this effort can help keep your conscience clean. It's a straightforward, hard-to-miss exercise that has the benefit of providing lunch (if you cook the lean bacon at the same time).

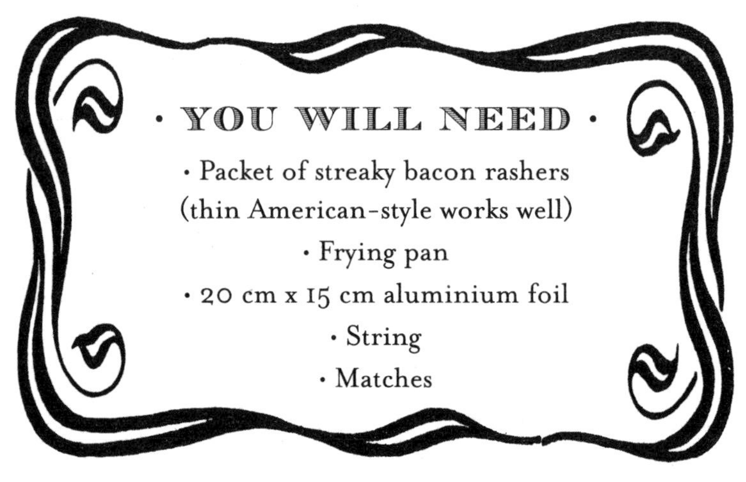

· YOU WILL NEED ·

- Packet of streaky bacon rashers (thin American-style works well)
- Frying pan
- 20 cm x 15 cm aluminium foil
- String
- Matches

METHOD

1. Cut the fat off the rashers and fry it until it becomes liquid.
2. As the fat is frying, form a tube from the aluminium foil, sealing one end of the tube and keeping the other open.
3. When the fat has completely melted, pour it into the tube.

4. Cut a length of string 2–3 cm longer than the height of the tube.

5. When the fat has begun to set, but is still malleable, feed the string into it so that the extra 2–3 cm protrudes (it will serve as a wick).

6. When the fat has hardened, take it from the tube and place on a non-flammable base (possibly also made from foil).

7. Light the fuse; it will burn like a candle but will emit a thick black smoke.

THE SCIENTIFIC EXCUSE

A car, without the right fuel mix (petrol and oxygen), will not 'flash burn' efficiently and will burn off the petrol with far more smoke. This experiment presents a mini-version of the same thing – a mix that scores high on fuel (the fat) and low on oxygen, providing the slow-burning part of things. The inherent impurities in animal fat (particles of meat and bone that burn black) mean that the smoke becomes very dark.

TAKE CARE!

Although the smokescreen isn't dangerous, it does produce a great deal of smoke. Now that, of course, is the whole point of the experiment but do plan ahead, taking into account ventilation, wind speed, who's likely to see or smell the smoke, etc.

MATCH ALERT!

This experiment involves the use of matches and should be conducted only by a responsible adult.

Potato Gun

This experiment takes the notion of a food fight and introduces a note of the 'arms race'. The basic ammunition is simple enough – the humble potato. But with a little preparation, and some help from Boyle's Law, the potato will 'go ballistic' – literally!

• YOU WILL NEED •

- 2 m metal or PVC tube (2–3 cm diameter)
- 2 potatoes
- Rubber stopper (optional)
- Goggles
- 2.5 m metal or wooden pole (diameter narrower than that of the tube)

METHOD

1. Press one end of the metal tube down onto a potato, so that a plug of potato becomes lodged at the end.
2. Repeat this process with the other end of the tube on another potato so that you have a potato plug at each end.

3. Hold the tube with one hand with one end pointing away from you – and away from anything breakable.

4. Position the pole at the closer end of the tube, just touching the potato plug. (If the pole is much narrower than the tube, add a rubber stopper to the end touching the potato.)

5. Put on your goggles and ram the pole quickly down the tube, pushing one potato plug towards the other.

6. The plug at the far ('barrel') end will shoot out at great force!

— THE SCIENTIFIC EXCUSE —

This experiment is an explosive demonstration of Boyle's Law – pressure increases as volume decreases. Scientifically stated, this means that: 'Under constant temperature, the volume of a gas is inversely proportional to the total amount of pressure applied.' In this experiment, the gas is the air inside the tube, lodged between the two potato plugs. When you shove the pole into the tube, you push one plug towards the other. This reduces the volume of the gas suddenly, thereby increasing its pressure suddenly. Something has to give: the burst of pressure sends the potato plug flying.

— TAKE CARE! —

Don't point the potato gun in the direction of anyone watching – or towards any breakable china or crockery!

Frankenstein's HAND

Here's a 'hands-on' demonstration of a scientific principle that has cropped up elsewhere in this book: the chemical reaction between a common acid and an equally familiar base. You can give this experiment something of a Halloween flavour by marking the glove with bones, veins and screws. Young children will love the special effects as the hand grows and grows.

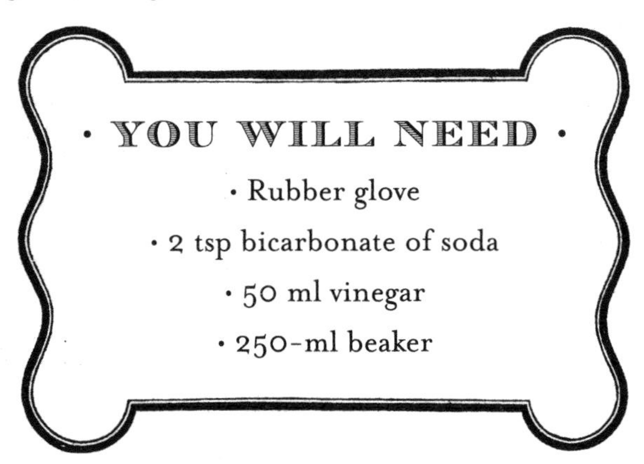

· YOU WILL NEED ·

- Rubber glove
- 2 tsp bicarbonate of soda
- 50 ml vinegar
- 250-ml beaker

METHOD

1. Pour the vinegar into the beaker.

2. Add the bicarbonate of soda to the inside of the glove. Hold the glove by its wrist and shake the powder into the fingers.

3. Carefully attach the glove to the top of the beaker so there's no gap.

4. Pull the glove upright by its fingertips and shake gently, allowing the bicarbonate of soda to drop into the beaker.

5. Stand back and watch as Frankenstein's hand begins to inflate.

— THE SCIENTIFIC EXCUSE —

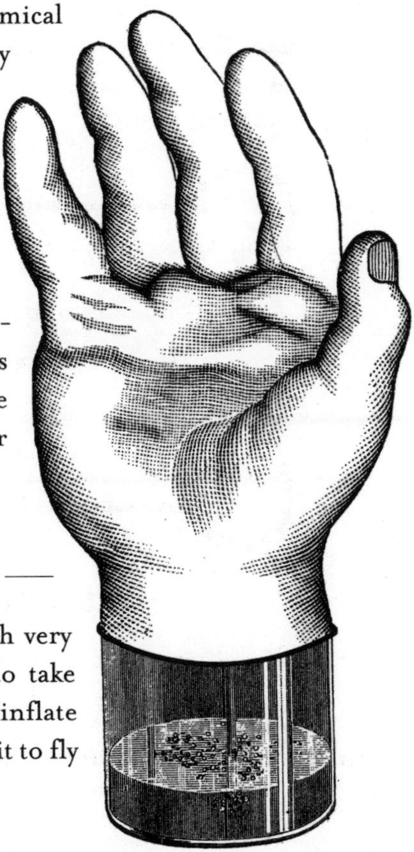

Bicarbonate of soda is a chemical base, which reacts strongly with the acetic acid of the vinegar. One of the by-products of this reaction is carbon dioxide, which increases the pressure inside the glove–beaker arrangement. As more gas is produced, this pressure increases further and pushes out the weaker surface (the rubber glove), inflating it gently.

— TAKE CARE! —

This is a safe experiment with very little risk. You might want to take care that the glove doesn't inflate too much, which could cause it to fly off the beaker.

Egg bungee jump

Have you ever done a bungee jump? Does it give you the shakes just thinking about it? How about trying a home version of it, but with an added, possibly very messy ingredient that might leave you quite literally with egg on your face.

YOU WILL NEED

- 6 eggs (although one should be enough)
- Pair of old tights
- Strong tape
- Scissors
- Ruler
- Supply of pennies
- Ladder (if no tree branch is available)
- Old newspaper

METHOD

1. Choose a spot for the bungee jump: ideally a tree branch outside but it could also work from a ladder. You want the egg to fall to within 3 cm of the ground but no closer; use the ruler to measure this distance.

2. Work out the weight of the egg beforehand to do a test run: hold the egg in one hand and keep adding pennies to the other until you feel the pile weighs the same as the egg.

3. Add the 'egg's worth' of pennies to a leg of the tights and tape the other end to branch or ladder.

4. Do a test run – let the coins fall and check their distance from the ground. Make any necessary adjustments to where you have tied the tights to the ladder or branch.

5. Now try the jump with the egg instead of the coins.

— THE SCIENTIFIC EXCUSE —

The nylon or other fabric of the tights has a natural elasticity – a certain force will take it so far before the strength of the fabric pulls it back. The measuring and the test runs might all seem like good fun, but you – or your assistants – are actually working out the components of the famous physics equation:

Force = Mass x Acceleration

Well done, Newton!

— TAKE CARE! —

If, for some reason, you're doing this inside then make sure you spread out some old newspaper for a landing site. And if you're still worried stiff about the mess – or if you've got the evidence of some 'near misses' on the floor – you could always do the experiment with a golf ball or a stone.

Turning Milk to Stone

There's something almost biblical in the title of this experiment, but once you've done it and understood what's happened you'll also be reminded of Little Miss Muffet. Regardless of which reference you prefer (maybe none at all), this demonstration scores high on the 'kewl' scale for youngsters.

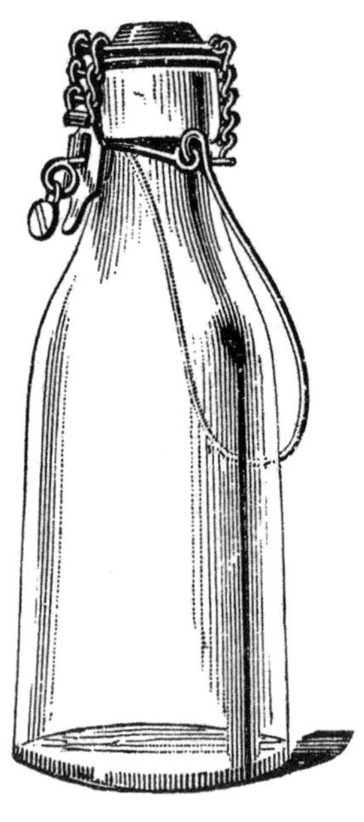

· YOU WILL NEED ·

- 350 ml skimmed milk
- 4 tsp vinegar
- Microwave
- Microwave-proof mixing bowl
- Strainer

· METHOD ·

1. Pour the milk into the mixing bowl.

2. Add the vinegar.

3. Put this mixture in the microwave and cook for one minute.

4. Remove from the microwave to find it now composed of a solid and a liquid.

5. Strain off the liquid.

6. Let the solid cool and then form it into shapes, which harden as they cool.

· THE SCIENTIFIC EXCUSE ·

The acid in the vinegar separates the curd (the semi-solid element) from the whey (the liquid) in the milk. The protein in the curd accounts for its rubbery quality – in fact, some of the earliest plastics were produced in a variation of this experiment.

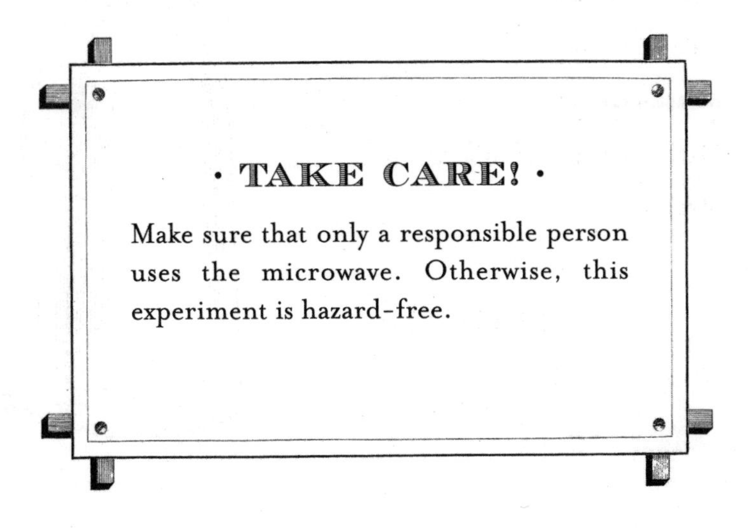

· TAKE CARE! ·

Make sure that only a responsible person uses the microwave. Otherwise, this experiment is hazard-free.

Iron Breakfast Cereal

No wonder it was so crunchy! Actually, cereal manufacturers have long played a part in providing us with essential minerals with our daily breakfast. Maybe this experiment could be a springboard to more lucrative exploits: 'There's gold in them thar cornflakes!'

YOU WILL NEED

- Single portion of 'fortified' breakfast cereal (check ingredients for iron)
- Magnetic tape
- Tongue depressor
- Non-metal mixing bowl (500 ml minimum capacity)
- 2 zip-lock sandwich bags
- Water

METHOD

1. Attach the magnetic tape to one end of the tongue depressor and seal it in one of the two sandwich bags.

2. Put the cereal in the other sandwich bag and crush it.

3. Pour this crushed cereal into the mixing bowl and cover with water.

4. Keeping it within the sandwich bag, use the now-magnetic tongue depressor to stir the cereal for 10 minutes.

5. Remove this stirring device from the mixing bowl and observe the tiny metal filings on the outside of it.

THE SCIENTIFIC EXCUSE

Humans need iron and other minerals in their daily diet. Many breakfast cereals are 'fortified' to help us meet this dietary requirement. Normally we wouldn't give this an extra thought, but after seeing the results of the experiment it all seems a little clearer – if perhaps a little less tasty. These iron filings are oxidised in the stomach and absorbed in the small intestine. If all the iron in your body was extracted, you'd have enough to make a couple of small nails.

TAKE CARE!

There's no risk with this experiment, but be patient in the stirring. Ten minutes is a long time but we're talking about small amounts of iron and any less stirring wouldn't be successful.

The Electric Spoon

We've all had that nightmare feeling when we've been following a recipe: 'Oh dear! I've mixed the ingredients too soon: they were meant to be separate at this stage.' Imagine if you could turn back the clock and pick the mixture apart bit by bit. Imagine further that you could do this with salt and pepper. Impossible? Read on and think again!

YOU WILL NEED

- 1 tsp table salt
- 1 tsp ground black pepper
- Plastic spoon
- Woollen sock (your hair will do if there's no sock to hand)
- Piece of paper (optional)

METHOD

1. Mix the salt and pepper on a smooth, dry surface.

2. Rub the spoon vigorously against the woollen sock (or your hair).

3. Slowly and very carefully lower the spoon towards the salt–pepper mixture.

4. A few centimetres from the surface, the pepper will jump up, grain by grain, to the spoon's surface.

5. Stop the experiment here if you simply want to show how it works. To complete the separation, continue to steps 6 and 7.

6. Slide a piece of paper under the spoon and shake the pepper grains on to it.

7. Rub the spoon against the wool again and repeat until all the pepper has 'jumped' to the spoon and only salt remains.

— THE SCIENTIFIC EXCUSE —

Several experiments in this book – including this one – depend on building an electrical flow by rubbing plastic or other substances against wool. As the negatively charged plastic spoon is lowered, it charges the salt and pepper by a process called induction. The nearest surface of the salt and pepper gain a positive charge and the far side a negative. Since 'opposites attract', the grains of pepper and salt jump towards the spoon.

TAKE CARE!

The pepper jumps up first because it's lighter than the salt. That's why lowering the spoon is so important: too fast and the salt and pepper would jump up together. If you want to complete the 'tidying up' after all the pepper has been removed, simply carry on with the process, moving the spoon a bit lower.

'Kick-start' Ice Cream

Most irresponsible experiments trigger satisfying bangs, smells, oozes and explosions. Not many produce anything that you'd like to eat. This experiment is an exception. It's fun, reckless, runs the risk of being very, very messy – but produces some of the best home-made ice cream that you'll ever taste.

You Will Need

- 250 ml double cream
- 250 ml whole milk
- 70 g white sugar
- 200 g rock salt
- Large bag of ice (available from most supermarkets)
- Empty coffee tin and its plastic lid (or a lidded plastic container of the same size)
- 15-litre plastic bucket with lid
- 2 tbsp vanilla extract
- A dessert spoon (more than one if you are tempted to share the result)

Method

1. Mix the cream, milk, sugar and vanilla in the coffee tin. It should be no higher than halfway up the tin.

2. Put the lid on the coffee tin and put it at the base of the large bucket.

3. Add a 10 cm layer of ice to the bucket in the space around the coffee tin, and then add the rock salt in the same area.

4. Fill the rest of the bucket with ice and affix the plastic lid.

5. Now, get your helpers to kick and roll the bucket around for about 10 minutes.

6. Open the bucket, remove and open the coffee tin. A layer of frozen ice cream will line the inside of the tin. Stir this into the chilled inner liquid to get the right consistency.

— THE SCIENTIFIC EXCUSE —

The key to the experiment is the salt, which lowers the freezing point of water. Kicking or rolling the bucket causes some of the ice to melt (the friction of knocking together). The water would normally refreeze, but the salt allows it to remain liquid despite being 'below freezing'. This superchilled water cools the coffee tin, and the ice-cream mixture begins to freeze. The kicking and rolling, however, have the same effect as stirring, preventing huge ice crystals from forming.

TAKE CARE!

Helpers have a habit of becoming overzealous during the 'kick and roll the bucket' phase of this experiment. It's probably a good idea to tape both lids (coffee tin and bucket) firmly in place beforehand.

Reinforced Rice

This wonderful, counter-intuitive experiment can be performed in a matter of a few seconds, especially if you keep your rice in the right sort of jar. There's something primal and moving about it – which might explain why some unscrupulous sorcerers in India have claimed magic powers when performing this trick.

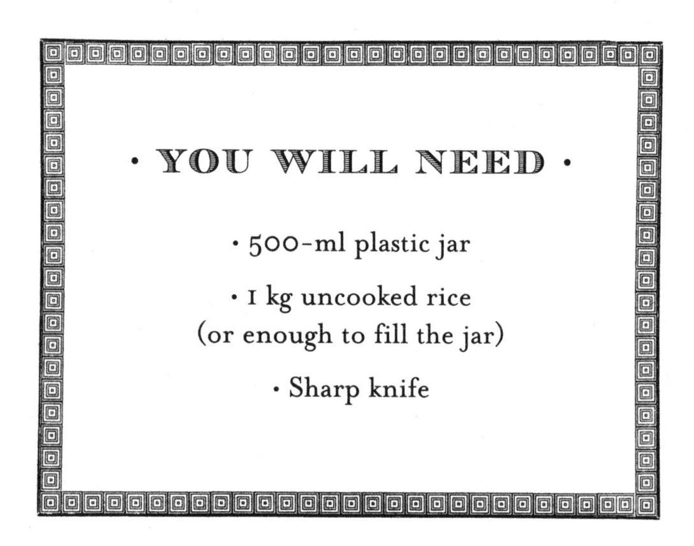

· YOU WILL NEED ·

- 500-ml plastic jar
- 1 kg uncooked rice (or enough to fill the jar)
- Sharp knife

· METHOD ·

1. Fill the jar with rice.

2. Repeatedly plunge the knife halfway into the rice, preferably at an angle.

3. Jab the knife fully into the rice.

4. Pull straight up on the knife and it will carry the jar of rice with it.

· THE SCIENTIFIC · EXCUSE

The force of pressure of thousands of grains of rice can overcome the force of gravity. Those preliminary 'stabs' of the rice served to settle it in tightly. The knife is sharp enough to be jabbed into this network of pressure, but the rice settles back in from all sides to hold the knife in place.

· TAKE CARE! ·

Only an adult should get involved with the knife-wielding. Observers should stand well back. Otherwise, this experiment is safe.

Eggs in the Nude

TV chefs have always relied on the 'here's one I made earlier' line to take into account lengthy preparation time. If you're planning on demonstrating this experiment to an audience, you might want to adopt the same approach, since a good two days of preparation is needed. (And it takes another day or two to get the real 'payoff'.) The experiment calls for two eggs, although only one is really needed. The extra one is held in reserve, since these 'eggs in the nude' are very delicate without their shells.

You Will Need

- 2 eggs
- 1 litre vinegar
- 150 ml corn syrup
- Wooden spoon
- Coffee mug
- 1-litre plastic mixing bowl with lid

Method

1. Put both eggs in a mixing bowl. Cover with vinegar, put the lid on the bowl and put it in the refrigerator.

2. After about 24 hours, carefully drain the vinegar (the shells will seem mostly eaten away at this point). Replace with fresh vinegar for another 24-hour bout.

3. At this point (48 hours from the start), remove the eggs with the wooden spoon, rinse them and put carefully to one

side. Discard the vinegar.

4. Show your audience one of the eggs, holding it carefully with the wooden spoon, and place it in the coffee mug. Then cover it with corn syrup.

5. After another 24 hours, extract the egg and notice how flabby it has become.

6. Now discard the corn syrup, rinse the mug and put the egg back in, covered this time with water.

7. After 24 hours (48 hours from Step 3 and four days from the start) you'll see that the egg is looking healthier again.

— THE SCIENTIFIC EXCUSE —

There are a few things going on here, all to do with chemistry. To remove the shell and make the eggs 'nude', you immersed them in vinegar. The acetic acid of the vinegar ate away the sodium calcium carbonate crystals of the shell, leaving only the rubbery membrane surrounding the egg. That membrane is slightly permeable, allowing water to move from an area of higher concentration (the egg white, for example, is about 90% water) to one of lower concentration (corn syrup is about 25% water). Replacing the corn syrup in the coffee mug with water reverses this process.

TAKE CARE!

None of the ingredients here is dangerous, so the main advice is one of taking great care in handling the eggs once the shell has been dissolved. The membrane holds things in ... just ... which is why we recommend having the egg in reserve.

Marshmallows on Steroids?

Just in case you've spent four days on the last experiment seeing to your 'eggs in the nude', we thought you'd like a real quickie. This experiment is probably one of the quickest and easiest in the book, provided you have a vacuum storage jar (often sold for coffee). And the best thing about it is that you don't lose out in the 'oomph' stakes. You'll never be able to read 'dramatic reductions' again with a straight face.

- 1-litre storage jar with vacuum pump
- Marshmallows (full-sized fluffy variety)

METHOD

1. Fill the jar nearly full of marshmallows.

2. Cover the jar and pump it to remove as much air as you can.

3. Examine the marshmallows, which should seem puffed up well beyond their original size.

4. Release the pump and observe as the marshmallows return to their original size.

— THE SCIENTIFIC EXCUSE —

Easy, if still dramatic and amusing. Pumping the air out of the container reduces the air pressure inside. When you consider that marshmallows are mainly air bubbles, with some linking solids, you can see how the air in the bubbles would expand to fill the less pressurised air around them. The marshmallow 'solids' (the non-bubble bits) are elastic enough to expand with the bubbles. When you release the pump, the marshmallows shrink back to normal.

• TAKE CARE! •

What can go wrong with marshmallows, for goodness' sake?

Vanishing Milk

Several of the experiments in this book are dramatic enough to qualify as magic tricks. This is one of them. Your audience will be dumbfounded as you pour a healthy glass of milk and then make to pour it back into the jug, undrunk. But it's gone! What happened? Is science to blame?

METHOD

1. Wearing gloves, cut open the centre of a disposable nappy with a sharp knife.

2. Scoop out 1/2 tsp of the powder inside the nappy (known as sodium polyacrylate) and put it in the drinking glass.

3. Pour the milk into the jug. You are now ready to perform the trick for your public.

4. Hold empty glass and full jug out before you, one in each hand.

5. Pour some milk into the glass.

6. Tell your audience that you didn't want the milk, and that you will pour it back. Try to pour the milk from the glass back into the jug. None will come out!

THE SCIENTIFIC EXCUSE

Sodium polyacrylate, the 'active ingredient' in most disposable nappies, is an especially absorbent substance. One test has shown that it can absorb 800 times its own weight in water, which makes up most of the milk. The sodium polyacrylate and milk mix to form a gel, which sticks to the base of the glass. When you turn it over to pour the milk back in, nothing comes out.

TAKE CARE!

Sodium polyacrylate can irritate the eyes and nostrils, so take care when you are dealing with it and also wear gloves. As for practical advice, this experiment works better as a trick if you don't use a clear glass. That way, your audience won't have a chance to notice the milky gel left at the glass bottom.

Straw through a Potato?

Not many of us have experienced the savage force of a tornado. Its 500-kph winds are strong enough to drive a piece of straw clear through a telephone pole. So you can imagine that if you left a drinking straw near a potato as a tornado approached, you might return to find the potato skewered by the straw. But can you imagine getting the same result by hand? Read on, and see how you can do it.

You Will Need

- Uncooked potato
- Rigid (non-bendy) straw
- Gardening glove (left if you're right-handed or vice-versa)

Method

1. Put the glove on one hand.

2. Hold the potato with the gloved hand, pinching it between thumb and index finger.

3. Hold the potato steady, pick up the straw (holding it in

the middle) and line it up with the potato.

4. Draw the straw back slowly, then stab the straw quickly into the potato.

5. If you're quick enough – and the straw is strong enough – you'll stab right through the potato.

— THE SCIENTIFIC EXCUSE —

The cylinder shape of the straw gives it surprising strength along through its length, although it remains weak and flexible crossways. That strength, coupled with the narrowness and sharpness of its edge, gives the straw a good chance of making it through the potato with ease.

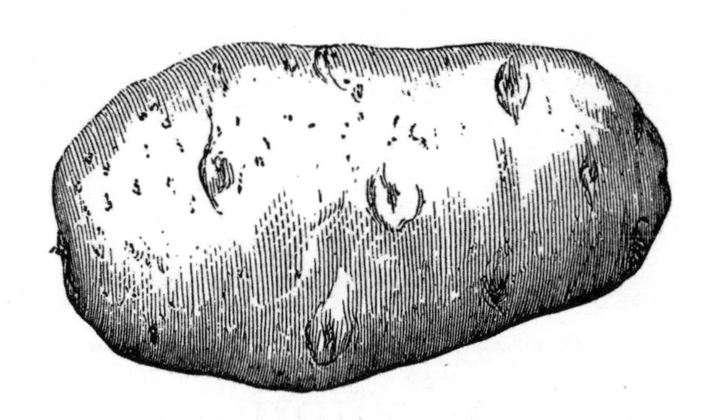

TAKE CARE!

This is a relatively low-risk experiment, although the same combination of strength and sharpness of the straw could lead to a hand injury if your aim isn't up to scratch. (That's why the gardening glove is called for.)

Peas & Queues

This little gem of an experiment scores high on the 'scientific excuse' side of the equation, offering a real chance to demonstrate – or learn – more about plants. Well, there's the justification out of the way. What the experiment also offers is a chance to capitalise on the spooky 'plip-plop' that others (those who aren't in on the experiment) will find so puzzling and unsettling. Try different positions in the kitchen – or elsewhere – to hide the set-up; see whether you can find somewhere that resonates. You can also try adding a bit of scrunched-up foil to the baking sheet to add a bit of audio variety. With their seeming patience, the peas appear to be queuing up to drop out of the glass.

You Will Need

- Wine glass (preferably bulb-shaped red-wine glass)
- Thin metal baking sheet or lid (about 20–25 cm wide, either square or circular)
- Dried peas (enough to fill the wine glass)
- Water
- Wide-mouthed (ca 6 cm) drinking glass with a heavy base

Method

1. Fill the wine glass to overflowing with dried peas.
2. Add water right to the brim of the wine glass

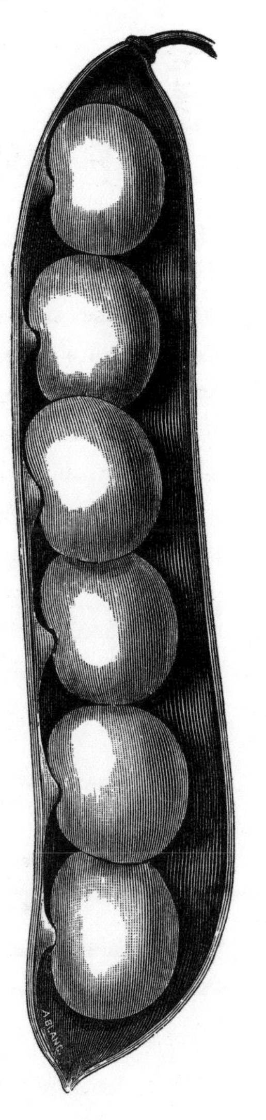

3. Place the baking sheet or lid on the stout drinking glass.

4. Put the wine glass on the baking sheet or lid.

5. Slowly, and over the course of the next few hours, the peas will fill with water and begin falling – with a spooky 'plink' – on to the baking sheet or lid.

— THE SCIENTIFIC — EXCUSE

This experiment is an ideal introduction to osmosis, the process by which water moves from an area of high concentration (the glass) to a lower one (the dried peas). An eager child can get a better grip on how plants draw water upwards from the soil. In this demonstration, though, each pea swells up as it draws in more water. There's literally no more room for these peas, which then tumble out of their position in the wine glass. Holding the baking sheet (on the drinking glass) amplifies each 'plink'.

TAKE CARE!

Make sure you don't leave this arrangement too close to the edge of a counter or on an unsteady surface.

Egg in a Bottle

Here's another take on the scientific principle of creating a vacuum. In this case, it's a hard-boiled egg that gets sucked into a bottle when matches go out. If this is all getting a bit easy, then try the raw-egg variation (see step 5 in the Method).

YOU WILL NEED

- Hard-boiled egg (peeled)
- Matches (long kitchen matches are best)
- 1-pint milk bottle
- Uncooked egg (optional)
- 200 ml vinegar (optional)

METHOD

1. Have a couple of practice runs resting the egg (narrow side down) on the lip of the milk bottle.

2. Light a match and then drop it into the bottle.

3. When the match has almost burned out (judge by the length of the unburnt remainder) place the egg back on the top of the bottle.

4. Watch as the egg seems to be sucked into the bottle.

5. Some people have managed to get the same effect with an uncooked egg which they had soaked for a few hours in vinegar. (The vinegar softens the shell.)

THE SCIENTIFIC EXCUSE

The explanation is pretty simple: the match warms the air, which then expands (some of it escaping) and becomes less dense. Putting the egg on the bottle snuffs out the fire because there is no oxygen supply left. The egg rests between normal air pressure (pushing it down) and weaker air pressure from inside (pushing up against it). No contest – the downward pressure wins the day.

TAKE CARE!

Make sure you do leave the match burning for as long as possible before putting the egg on the bottle. The longer the match burns, the hotter the air inside becomes ... and therefore the greater the difference between the air pressures inside and out.

MATCH ALERT!

This experiment involves the use of matches and should be conducted only by a responsible adult.

Sandwich in a Jar

You know how it is: sometimes you just don't like the bread in your sandwich but you love the gooey filling. Wouldn't it be great to be able to get rid of that stale old bread on either side and just concentrate on three spreadable layers? You could even store it in a jar!

YOU WILL NEED

- A narrow glass jar with a lid or cork
- 3 tbsp water
- 3 tbsp cooking oil
- 3 tbsp honey, treacle or golden syrup

METHOD

Pour the three liquids (water, oil and whichever of the last three you chose) into the jar. Then cover the jar. Within a few minutes, the 'sandwich' will form before your eyes: the liquids separate into three distinct layers.

THE SCIENTIFIC EXCUSE

This is all about density (the amount something weighs in a particular volume). The volume remains the same (3 tbsp) in this experiment, so the densest liquid (the honey, treacle or golden syrup) settles at the bottom. The oil is denser than the water, so it settles in the middle, leaving the clear water layer at the top.

TAKE CARE!

You'd better not take the sandwich idea too far and make anyone eat it!

Supersalt!

We cook with it, sprinkle it on chips and popcorn and sometimes preserve food with it. But using salt as a front-line defence seems to be stretching a point. Or does it? Maybe there's more to those little crystals than we ever imagined …

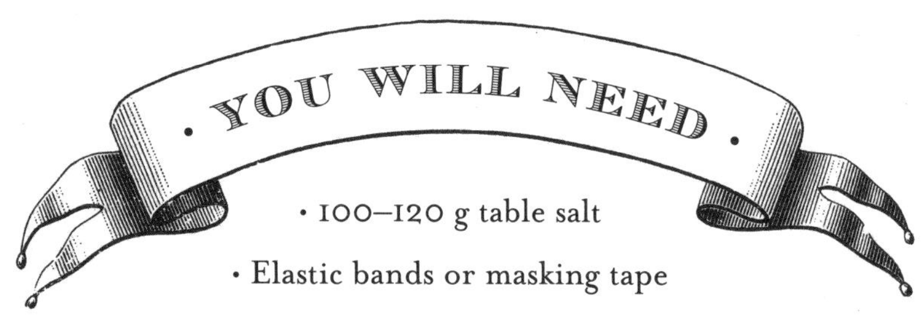

- 100–120 g table salt
- Elastic bands or masking tape
- Cardboard tube (e.g. from paper towel or cooking foil)
- Greaseproof paper
- Broomstick or wooden dowel (narrow enough to fit inside the cardboard tube)
- Scissors

METHOD

1. Cut two squares of greaseproof paper (each 15 cm x 15cm).

2. Use an elastic band or masking tape to fix one of those squares to one end of the tube. (It should be spread taut over the end like the skin on a drum.)

3. Feed the broomstick or dowel into the other end and then push through – note how easily it breaks through the paper.

4. Remove the torn paper and replace it with the second piece you had cut.

5. Now fill the tube up with about 8–9 cm of salt.

6. Try plunging the broomstick through again. Wait for the salt to spray out ... but maybe it won't!

— THE SCIENTIFIC EXCUSE —

This experiment is all about the direction of force. With no salt to block it, the broomstick will easily pierce the grease-proof paper at the end of the tube. The addition of the salt, however, changes the rules. The thousands of crystals deflect the force, channelling it away from its original course and spreading it out towards the edge of the tube – rather than towards the thin paper at the end.

TAKE CARE!

'Everything in moderation', as they say about so much in life. Someone trying to recreate the force of a Channel Tunnel train might succeed in breaching the salt defence, but the whole point here isn't to test our strength, but to observe the strength of the salt.

Carrot First Aid

Wouldn't it be wonderful to be able to give the 'kiss of life' to wilted, droopy vegetables? This experiment gives you the chance to do just that, giving a limp carrot the chance to 'stand up and be counted' once more – or at least to feel a little more solid.

• You Will Need •

- Good-sized carrot which has become limp
- 300-ml drinking glass (clear)
- Small cup
- Clear plastic straw (cut to a 3-cm length)
- 4 toothpicks
- 1 tsp water
- 1 tsp of sugar
- Plasticine

Method

1. Fill the glass about 3/4 full of water.

2. Hollow out the top of the carrot (about 1 cm square and 1–1.5 cm deep).

3. Set the length of straw into this hole and seal the edge with Plasticine.

4. Stick the toothpicks at even distances into the sides of the carrot near its top and rest these on the rim of the glass, with the carrot mainly submerged.

5. Mix the sugar with the equal amount of water in the small cup and fill the straw (which juts out from the carrot) about halfway up.

6. Observe in an hour or two – the water level in the straw should have risen.

— THE SCIENTIFIC EXCUSE —

This experiment is all about the process called osmosis. Water passes through plant-cell walls as it moves from an area of high concentration (of water) to an area of lower concentration. In this case, the carrot has become limp because it has lost much of the water that gave it strength. While in the glass, it absorbs water from its surroundings, becoming more rigid again. In a growing carrot, some of this water would be channelled upwards through the top of the orange bit; the addition of the straw (as a sort of gauge) shows how this water level rises.

TAKE CARE!

There's no danger inherent in this experiment. Make sure, though, you use a sugar solution in the straw: ordinary water might evaporate.

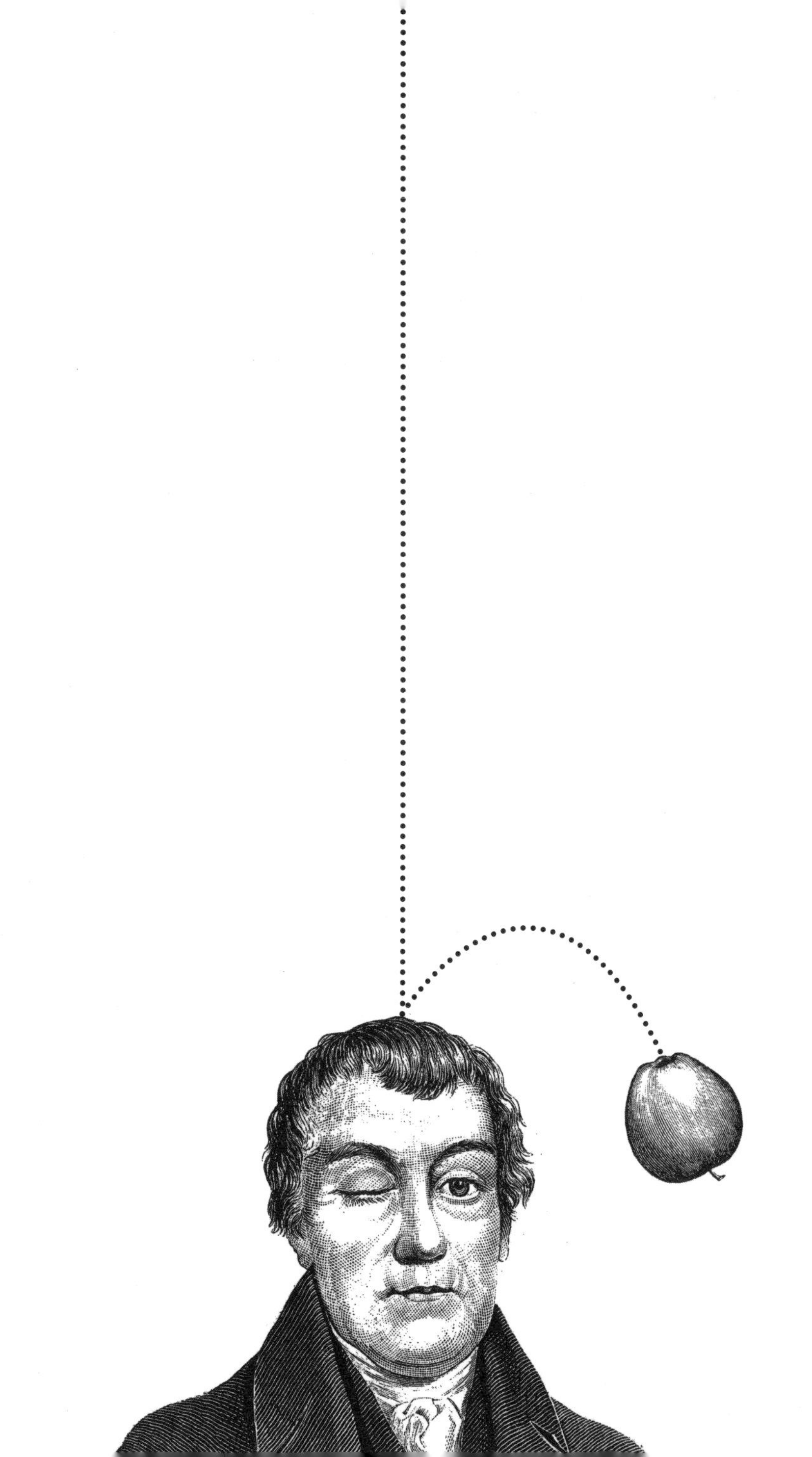

How Moving!

Can you remember Sir Isaac Newton's First Law of Motion? His Second? His Third? Don't worry if you can't, because you'll be demonstrating some of them yourself if you carry out some of the following experiments. Everyone loves a chance to build something that rolls, flies, hovers or blasts off. And to do that in the cause of science – with an eager audience cheering you on – well, that's a little bit of heaven.

Anti-Gravity Water

We're all aware – even unconsciously – of the forces of motion that govern the world of physics around us. Most of us have slid towards the side of a car as it rounded a bend or even, as children, leaned into a curve as we sped on our bikes. Here's a chance to get to the root of that 'force of nature'. Give it a whirl. Cheers!

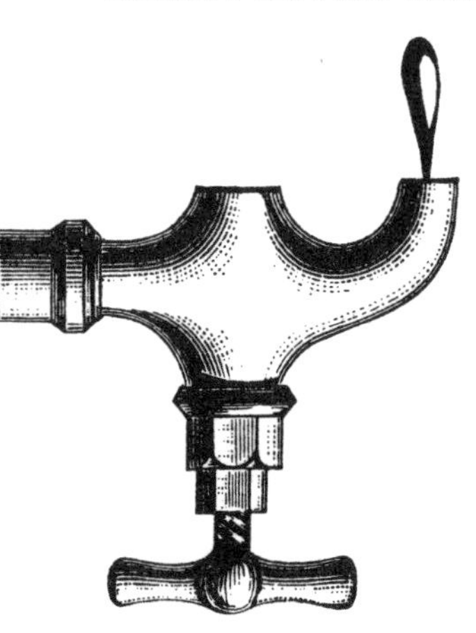

· You Will Need ·

- 18 cm x 18 cm square of stout card
- 2 x 1-m string
- Hole punch
- Plastic or polystyrene cup
- Some coins (optional)
- Water

Method

1. Punch holes about 2 cm in from each corner of the card.

2. Feed one end of a piece of string up through one of the holes, and the other piece of the same string up through the hole diagonally opposite.

3. Repeat step 2 with the other piece of string so that you have formed a string 'X' at the base of the card. Lay the card somewhere flat.

4. Ensure that the four upward-leading strings are the same

length, then tie them so that you can pick up the card 'tray' evenly by holding the knot and lifting.

5. Half-fill the cup with water and put it in the centre of the tray.

6. Holding the knot, lift the tray and then whirl it round. None of the water should spill out, even if you whirl it in a vertical 'orbit'.

7. With the whirling still vertical, slow the speed down carefully and pluck the cup from the tray before it falls off. If you've got the knack, none of the water will have spilled out in the experiment.

— THE SCIENTIFIC EXCUSE —

You have just been demonstrating Newton's First Law of Motion: 'An object in motion will remain in motion unless acted upon by an external and unbalanced force.' In this case, swinging the cup sends its contents outwards. The tension of the string exerts the 'external force', causing the cup to move in a circular path. The water, however, continues to try to move in a linear (straight) path, so it is pushed against the base of the cup. The inward movement of the cup is called centripetal force.

TAKE CARE!

The risks here are pretty obvious, so you should do this experiment outside, where there's no need for a clean-up or public hearing if things go a little wrong. And if you're really in the mood to tone down the irresponsibility even further, try using a few coins in the cup instead of water.

Floating Away

'It's all just smoke and mirrors.' This expression, once confined to the music-hall critic's assessment of conjurors' performances, has entered the language to imply any trickery that leaves audiences confused or bemused. Well, this book has helped readers use – or produce – smoke to get some dramatic results. Now how about a little mirror work? It's a bit time-consuming setting it up, but you'll be able to use the set-up again and again. This experiment (or trick) is best done with an audience, so keep them mum about the preparations and secret.

You Will Need

12 lengths of wood (75 cm x 2 cm x 2 cm)

Panel pins or wood adhesive

1 m x 1 m square mirror

Method

1. Construct a wooden cube by joining the 12 lengths of wood held together with the panel pins or wood adhesive.

2. Find a table that will hold your weight and will have enough room to hold the cube. Position a chair near it (so you can climb onto the table easily), all the while ensuring

that your audience is in another room – or behind a curtain, if you're feeling theatrical.

3. Put the cube on the table, then slide the mirror straight down into it. The mirror should be upright (mirror side facing your audience) and positioned diagonally when seen from above.

4. Mount the table and step into the cube, straddling the mirror. Stand still.

5. Invite your audience in, or draw the curtain.

6. Begin a series of slow movements with the leg that's in front of the mirror, keeping the other leg firmly in place for support.

7. The mirror will make it look as though both your legs are leaving the table, so capitalise on this optical illusion by pretending to float away, swim in mid-air, etc.

THE SCIENTIFIC EXCUSE

This is, of course, all down to the optical illusion created by the mirror. In strict scientific terms, light travels from your visible leg to the mirror and is reflected outwards, making people perceive it as your 'other' leg – which, of course, is firmly on the ground.

TAKE CARE!

Does this book need to remind you about the risks involved in breaking mirrors? Make sure the mirror is reasonably secure as it rests against the wooden cube. Any play in the mirror would ruin the effect.

Film canister rocket

Harness the powers of chemistry and call in the aid of Sir Isaac Newton to send your home-made rocket ... well, not quite into orbit but at least zooming off its kitchen launch pad. This experiment can be a group effort from start to finish, from the cutting of the fuselage to providing the voice for the 'final countdown'.

You will need

- Film canister with click-shut lid
- Coloured paper (lightweight)
- Several sheets of toilet tissue
- 3 tsp lemon juice per flight
- 1 tsp bicarbonate of soda per flight
- Scissors
- Sticky tape
- Glue
- Teaspoon

Method

1. Cut a piece of coloured paper to about 12 cm x 6 cm: this piece will become the fuselage.

2. Line the long end of this piece of paper along the open end of the canister and tape it on lengthways, keeping the open end clear of any tape.

3. Roll this paper around the canister (it overlaps a bit) and tape it into a secure cylinder. (One end of the paper cylinder should be flush with the open end of the canister and the other end extends past the closed end.)

4. Cut a circle of 4 cm diameter from more coloured paper, and then cut a pie-shaped segment that's about 1/8 of the circle. Tape the larger piece that you are left with shut: this will be the nose cone.

5. Glue the nose cone to the end of the cylinder (be sparing with the glue, to minimise weight).

6. Cut three triangles to form old-fashioned 'fins' for the base of the rocket. Fold each triangle slightly to create a narrow flap. Glue each flap at equal distances on the sides at the base of the rocket.

7. Hold the rocket upside-down and add the lemon juice.

8. Stretch a piece of toilet tissue over the open end of the upturned rocket. Pinch it to the edges of the rocket so that it is held lightly across the opening (with some slack – but not touching the lemon juice).

9. Pour 1 tsp of bicarbonate of soda into the slack and – still holding the rocket upside-down – replace the lid. Trim off any excess paper.

10. Turn the rocket over slowly and place it on the prepared launch site. Stand back and wait: take-off is usually within 20 seconds.

THE SCIENTIFIC EXCUSE

The reaction between the lemon juice (an acid) and the bicarbonate of soda (a base) produces carbon dioxide. This gas builds up pressure inside the canister until it blows the lid off the bottom. That's where Newton's Third Law of Motion ('For every action there is an opposite and equal reaction') comes into play. The canister is blown upwards with an equivalent force to the one that blew the lid down. Lift-off!

TAKE CARE!

Make sure you your launch site is on a waterproof (or water-resistant) tablecloth or a floor that won't be stained easily. The timing of the launch (once the ingredients are inside) can be a little hard to predict, so take care not to be directly above it. The chemistry behind the propulsion is fail-safe, so if your rocket fails to launch, examine it to see whether it's too heavy. If so, modify it by reducing the size of its components or the amount of tape or glue used: see if it works now.

Tennis Ball Moon Bounce

Can physics really be this much fun? How in the world can you manage to bounce the tennis ball that high? Expect physics faculties to be stormed by applicants in the next few years after you've given this demonstration a few times. You'll need three things to make the experiment work well: a steady hand, a hard surface (for a good bounce) and space. It's best performed on a driveway or in a car park.

YOU WILL NEED

- Tennis ball
- Basketball (a football or volleyball will work, but not as dramatically)
- Golf ball (optional)
- Table-tennis ball (optional)
- Cardboard tube (optional)

METHOD

1. Hold the basketball out at chest height and drop it: note how high it bounces.

2. Repeat with the tennis ball.

3. Now hold the two together at arm's length, with the ten-

nis ball above the basketball and touching it on the top (at the 'north pole' of the basketball).

4. Drop the pair and note how high the tennis ball bounces.

5. If it's hard to get the alignment right, try using a golf ball beneath a table-tennis ball.

6. And if it's still difficult, hold the golf ball and table-tennis ball inside a cardboard tube (to align them with each other and vertically): this should make it easier to perform.

— THE SCIENTIFIC EXCUSE —

This is a wonderful demonstration of Conservation of Momentum. Momentum can be expressed as mass times velocity. When the balls collide, their momentums are equal, but because the mass of the basketball is so much greater, the tennis ball acquires a much greater velocity to balance the equation. The result: it shoots off to space. The same principle applies to a cricket bat (massive) hitting a ball (less massive): the ball goes sailing for a six while the bat remains in the batsman's hands.

TAKE CARE!

Provided you've given yourself enough space to perform this experiment (i.e. outside), there should be no concerns. Otherwise, we'd be looking at four sports that would face bans on the grounds of safety.

Dancing Mothballs

If the Mexicans can have jumping beans, then why can't you have dancing mothballs? Just a few easy steps and you'll see how easy it is. And how in the world is all of this irresponsible? Well, just look at those ingredients and imagine how your kitchen will smell after you mix them. How popular will you be then?

You Will Need

- A wide-mouthed glass jar or drinking glass
- Several mothballs
- 60 ml vinegar
- 10 ml bicarbonate of soda
- Water

Method

1. Fill the jar or glass with water, leaving 2 cm at the top.
2. Add vinegar and bicarbonate of soda and stir gently.
3. Add two or three mothballs.

4. Watch as the mothballs slowly 'dance' – sinking first, then slowly rising, then sinking again.

THE · SCIENTIFIC · EXCUSE

The vinegar, bicarbonate of soda and water react to release carbon dioxide. The mothballs seem smooth, but in fact have very rough and uneven surfaces. Carbon dioxide bubbles can lodge on these surfaces. When enough are attached, the mothball becomes less dense than the liquid and rises to the top. At the surface, much of the carbon dioxide is released into the surrounding air, making the mothball denser once more and allowing it to sink ... and repeat the process.

TAKE CARE!

This is a fairly 'low-risk' experiment. The biggest concern is discarding the liquid and its contents afterwards so that the next person in the kitchen isn't tempted to drink it!

The CD Hovercraft

It's everyone's dream to be able to glide around as if they were on a flying saucer. That type of friction-free movement does, in fact, lie at the heart of hovercraft technology. This experiment lets you use everyday objects as a gateway to this brave new world of transport. Be careful, though: get your answer straight before you hear: 'What have you done to my new Amy Winehouse CD?'

METHOD

1. Make a 10–15 mm hole in the centre of the cap.

2. Glue the cap to the centre of the CD or DVD, so that the holes align (and so that the flat of the cap is touching the CD).

3. Blow up the balloon and twist the end tight (but not tied).

4. Carefully roll the balloon's 'lip' over the edge of the cap.

5. Place the CD on a smooth surface and release.

6. The 'hovercraft' should slide easily along the surface after having had the slightest touch.

THE SCIENTIFIC EXCUSE

Air rushes out of the balloon through the hole to provide an 'air cushion' below the CD. This cushion supports the entire lower surface of the CD because air is escaping on all sides. And without the friction of rubbing against the surface (the air cushion has eliminated it), the 'hovercraft' can move freely. The same principles enable real hovercrafts to transport passengers and cars across open water.

TAKE CARE!

Be careful not to use too much glue when attaching the cap to the CD (you might block part of the hole). And, as mentioned already, don't use anyone's precious new CD!

Two-stage rocket

The NASA website points out how rocket technology is just a variation on balloon technology, so mastering a two-stage balloon rocket should put you in the same engineering elite who produced the NASA space shuttles and the proposed Mars missions. It's wonderful to think that some of the same scientific principles are at work in this common-or-garden demonstration that anyone can master. It's not rocket science after all, or is it?

You will need

- 2 volunteers
- 2 long, narrow party balloons (inflatable to around 60 cm)
- 25–30 m clear fishing line
- 2 x 3-cm lengths of drinking straw
- Masking tape
- Scissors
- Empty 1-litre plastic bottle

Method

1. Thread the two lengths of straw on the fishing line.
2. Tie the line taut, attaching it to two strong objects such as trees or clothes lines.
3. Cut a ring of plastic 2 cm wide from the middle of the plastic bottle.
4. Blow up one of the balloons and pinch its end shut; press this pinched end to the inside of the plastic ring. (This first balloon will become the second stage of the rocket.)

5. Continue holding the pinched end of the first balloon against the ring while you insert the second balloon part-way through the ring.

6. Now blow up the second balloon, eventually letting go of the first balloon's end when it has been pressed to the inside of the ring by the second balloon.

7. Pinch the end of the second balloon: you should now have a 'two-stage balloon' in your hands.

8. Have the second volunteer attach each balloon 'stage' to a straw section (which is free to slide along the fishing line). Continue pinching the end of the second balloon while you pull the combination back to the end of the line.

9. Release the balloon and it should speed off down the line, discarding the first stage (the second balloon you inflated) along the way.

— THE SCIENTIFIC EXCUSE —

'For every action, there is an opposite and equal reaction.' This principle, observed by Newton, lies at the heart of rocket (or in this case, balloon) science. Letting go of the balloon allows the high-pressure air to rush out into the less pressurised air around it. But this rushing out provokes an opposite force which propels the rocket forwards. The same process is repeated when the first stage is depleted and it's the turn of the second stage to push ahead.

TAKE CARE!

This experiment works best (the stages travel further) if the fishing line is very taut and also level. This reduces friction along the track.

SCIENTIARUM
FAVORE

A Lot of Hot Air

Some of the most amusing and magical experiments are the ones that call on an invisible ingredient – the air around us – to play a starring role. Many of the experiments in this chapter have that magical aura, putting air to work to make things float, stick and propel … or even to 'haunt' an empty bottle of vintage wine. Your audience will be eating from the palm of your hand when they get wind of these experiments.

– The –
HOVERING BALL

Have you ever phoned a hairdresser and tried to book yourself in for a 'Bernoulli'? No? Maybe you should, because hairdressers have a tool to demonstrate one of the most far-reaching principles of physics, first noted by Swiss scientist Daniel Bernoulli in the 18th century. Take 30 seconds to grasp the essence of Bernoulli's timeless observation. In the meantime, have a ball!

YOU WILL NEED

- Hand-held hairdryer
- Table-tennis ball
- Cardboard tube

METHOD

1. Set the hairdryer to a low temperature but high speed and point it upwards.

2. Lower the ball into the upward flow of air and let go.

3. Watch as the ball remains suspended.

4. With the ball still suspended, lower the cardboard tube down towards it slowly.

5. The ball will be sucked up into the tube.

— THE SCIENTIFIC EXCUSE —

This experiment, like some others in this book, works because of Bernoulli's principle (see p. 111). Bernoulli, noted that a gas or liquid loses pressure as it gains speed. In this case, the air moves from the hairdryer and around the ball, and this moving air has low pressure. The pressure remains unchanged (stronger) in the calmer air outside the column of moving air, plus in the air behind the ball. These forces keep the ball from moving too far to the left or right – or even upwards. Adding the tube from above channels the moving air into the narrower space. This makes the air gain speed – and lose pressure – causing the ball to follow it upwards.

TAKE CARE!

Because it uses an electrical appliance, this experiment should be carried out by a responsible person. Remember, it's the movement of the air – and not the heat – that makes the experiment work, so use the coolest setting on the hairdryer.

Viking Funeral

Viking warriors who died in battle would be sent to Valhalla in style, or at least that's how Hollywood has always imagined their culture. A longboat would be set ablaze and sent off to the open sea on its way to Valhalla. You can add a little Viking ornamentation to the basic structure of the boat in this experiment to book your own meeting with Odin, but the demonstration appeals to the child in all of us no matter how it's presented. The experiment could be done in a bath but ideally – and to benefit from the whole 'Viking funeral' angle – it should be performed on the calm waters of a small pond.

YOU WILL NEED

• Aluminium tube for tablets or pills (about 10 cm long, with a screw top)

• 4 small candles

• Matches

• An empty shallow tin (e.g. for sardines or pilchards) just large enough to hold the tube lengthways

• Drill or nail to puncture the screw top of the tube

METHOD

1. Using a drill or nail, make a small hole (from the inside out) in the screw top of the aluminium tube. The hole should be just off-centre so that steam will be able to escape from it later.

2. Half-fill the tube with water and screw the top back on.

3. Place the candles in the tin, in a line against one of the long edges. You can secure them in position with a little wax dripped into the tin.

4. Place the tube (with the hole on top) inside the tin against the other long edge, and secure it in the same way. The candles and tube should all fit quite snugly.

5. Light the candles and place the tin on the surface of the water.

6. The 'Viking longship' soon begins to travel under its own head of steam.

THE SCIENTIFIC EXCUSE

The candles alongside the tube heat the water inside. This water soon reaches boiling point and the steam needs to find an escape point: the small hole. The steam expands quickly and creates a noticeable recoil as it leaves the narrow opening.

TAKE CARE!

On a practical level, note that the vessel is quite fragile so it's best to wait for calm weather – and water conditions – before conducting the experiment outside.

MATCH ALERT!

This experiment involves the use of matches and should be conducted only by a responsible adult.

The Bold Little Ball

This experiment is a little gem. It has just two ingredients and can be performed in a matter of seconds. But irresponsible? Well, it's not messy, or loud or likely to give anyone an electric shock. But it could become dangerous if you dare someone else to blow the ball from the funnel. Then you might need to take cover from your purple-faced – and very frustrated – friend!

You Will Need

- Narrow-mouthed funnel (a cooking funnel is ideal)
- One table-tennis ball

Method

1. Put the ball into the funnel and hold upright.

2. Tilt your head back and put the mouth of the funnel to your lips.

3. Blow.

4. While blowing, slowly tilt the funnel so that it should 'pour out' the ball – yet it doesn't!

THE SCIENTIFIC EXCUSE

Here is where things get even better. The ball stays in place because of the Bernoulli Effect – the same scientific reason that a jumbo jet stays airborne. Basically, the Bernoulli Effect describes how moving air (in this case, the air rushing into the funnel and along its sides) has lower pressure than air that's still. A light object, such as a feather or the ball in this experiment, will move to the area of lower pressure: the air behind the ball – even when the funnel is pointing down – has higher pressure and keeps the ball in place.

TAKE CARE!

Luckily, there's no direct risk associated with this experiment. If you choose to dare people to try it – or even make a bet – then you might face some angry reactions ...

– The –
UNPOPPABLE BALLOON

Here's another one of those experiments that's so startling, so counter-intuitive, that you should really perform it as a magic trick. After all, how often have you seen – much less been able to perform yourself – a demonstration as baffling as this one? And if that weren't enough of a billing, it's really easy to do.

YOU WILL NEED

- Balloon
- 30-cm cooking skewer
- Cooking oil

METHOD

1. Inflate the balloon, but not quite fully. Tie the open end as you would normally.

2. Brush the skewer on both sides with cooking oil.

3. Hold the skewer with one hand and the balloon in the other (making sure that the knot of the balloon is on the balloon's 'equator' rather than on one of the 'poles'). If you're doing it as a trick, you could have an assistant hold the balloon.

4. Slowly pierce the balloon with the skewer, twisting slowly and taking care to enter at the knot and to exit exactly opposite (where there should still be a small amount of slack).

5. Equally carefully, remove the skewer; hold the balloon by pinching the far hole. The balloon will still not burst although you will notice it slowly losing air.

6. Puncture the balloon deliberately by quickly piercing the stretched side of the balloon.

THE SCIENTIFIC EXCUSE

The rubber of the balloon consists of many interlinked long molecules, which are known as polymers. The molecules are held together in a pattern called cross-linking, which allows for some elasticity. If there's still some slack in the cross-linking (as there is in the knotted end and its opposite extreme), the polymers can 'close ranks' around a small incision. The rubber actually forms a seal around the skewer that has just pierced it. Lubricating the skewer makes this entrance a little easier, and it helps provide a temporary patch when the skewer is removed. At the end of the experiment, you can burst the balloon easily by puncturing the stretched side, which hasn't enough slack to form a protective seal.

TAKE CARE!

Make sure an adult wields the skewer. If you do decide to perform this as a magic trick, don't let your audience see you perform step 2. Also, make sure you pop the balloon soon after the 'payoff' – before people notice that the balloon is slowly deflating anyway.

– The –
Clinking Claret

'Darling, I'm just finishing off that claret so that I can teach the children a little about air pressure.'

'Anything you say, dear. You don't think you're full of hot air yourself, do you?'

YOU WILL NEED:

- Empty wine bottle
- Coin

(the British 10p coin is ideal)

METHOD

1. Keep the wine bottle somewhere cool for a few hours, or preferably overnight (you could even scour your recycling bin outside!).

2. Take the cooled wine bottle inside, moisten the rim and put the coin on its open mouth.

3. Wrap your hands around the bottle.

4. Wait until the coin begins to move as if being fingered by an unseen hand.

— THE SCIENTIFIC EXCUSE —

The air inside the cooled bottle takes up relatively little space, but as it warms up (because of your hands) it tries to expand. This would-be expansion builds up pressure inside the bottle, but water tension holds the coin in place. Eventually the pressure is enough to overcome the water tension – briefly – allowing some of the air to escape. The coin pops open when this happens, then settles again when the air pressure inside has dropped again.

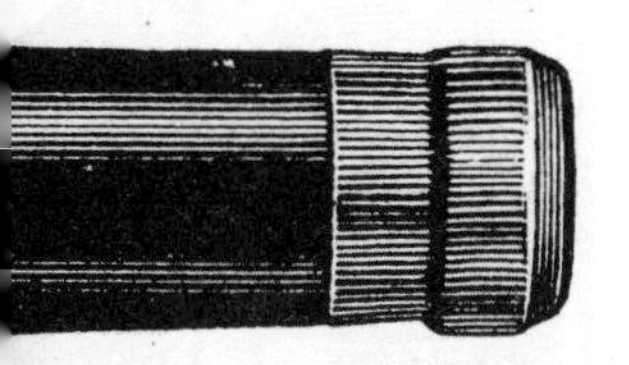

— TAKE CARE! —

The only risks you'll run with this experiment, as forewarned, are confined to emptying the bottle before embarking on the experiment.

– The –
TEABAG BALLOON

There's something magical about taking an everyday object and getting it to behave in a particularly unpredictable way. In this case the object is a humble teabag, which can be carefully taken apart and then set alight. The teabag will float upwards, as if by magic. Because even a slight puff of wind could disrupt this delicate experiment, it's best done inside – or outside in very sheltered surroundings on a calm day.

• YOU WILL NEED •

• Teabag (the sort with a string and tag)

• Matches

• Small plate

• Staple remover (or strong fingernails)

METHOD

1. Remove the string and tag from the teabag.

2. Using a staple remover or your fingernails, remove the staple from the teabag.

3. Unfold the teabag and pour the loose tea out of it.

4. Stand the emptied teabag – which is now shaped like a cylinder – on its end, on a small plate.

5. Light a match and set fire to the top of the teabag.

6. As the flame burns down, the burning teabag will shake a little and then float up into the air.

THE SCIENTIFIC EXCUSE

It's the old chestnut – 'hot air rises' – at work here. The fire causes two things to happen. One is that the air around the teabag gets hotter and hotter. Another is that – because of the burning – the teabag loses weight (or mass, if you want to be really scientific). These two factors converge, eventually reaching a point where the air is warm enough – and the teabag light enough – to produce lift-off.

MATCH ALERT!

This experiment involves the use of matches and should be conducted only by a responsible adult.

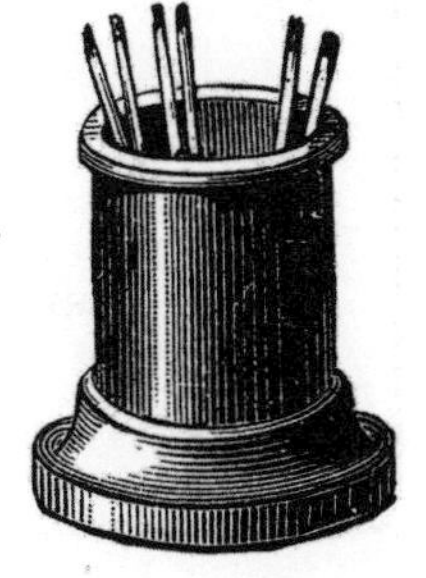

The last straw!

Some of the best science experiments are the ones that overturn our preconceptions, prompting calls of: 'That's not right – it can't be.' This experiment is a fine example of one of those myth-busters, bound to get someone hot under the collar. Getting a drinking straw to work is child's play, isn't it? Well, find me a child because this one's not working!

YOU WILL NEED

- Clean glass or plastic jar with a tightly fitting lid
- Drinking straw (plastic or paper, but plastic works better)
- Water
- Plasticine
- Hammer and nail or awl

METHOD

1. Fill the jar about 2/3 full of water.

2. Fix the lid on tightly.

3. Use the hammer and nail or awl to make a hole in the lid; work the hole until it's wide enough for the straw to fit through it.

4. Slide the straw downwards through the lid, judging how far down it should go by lining the lid up against the jar.

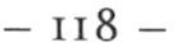

5. Seal the edge of the hole around the straw with some Plasticine, working it in so there's no gap at all.

6. Screw the lid back on the jar and try to drink through the straw. You – or your disgruntled volunteer – should find it impossible.

— THE SCIENTIFIC EXCUSE —

The preconception mentioned already is that we suck water directly when we drink through a straw. This experiment shows that it's all about air and air pressure. The air around us exerts pressure on everything it touches, including the surface of a drink in a glass or open jar. When you put a straw in a drink normally (that is, when the drink is in an uncovered container), the pressure of the air on the drink is the same as the pressure of the air coming through the straw. But when you suck on the straw, you remove some of the air and reduce the pressure in the air that's left in the straw. The air pressing down on the surface of the drink, however, still has the same pressure. The difference in pressure pushes the liquid into – and through – the straw. Sealing the lid tightly blocks the outside air from pressing down on the liquid in the jar, so now matter how much you suck, there's no pressure from inside to force the liquid into the straw.

TAKE CARE!

There's no danger involved with this experiment, but it's worth noting that plastic straws can be reused more often than paper ones since they are that much stronger.

Balloon Power

We often lose sight of the fact that air has mass and exerts pressure. Sure, we all register when the Met Office talks about air pressure, but that's a little abstract. This experiment brings it all home, though, and is something of a work in progress. Some children have managed to attach more than two dozen cups to their balloons – see how (and why that amount is so impressive) by reading on.

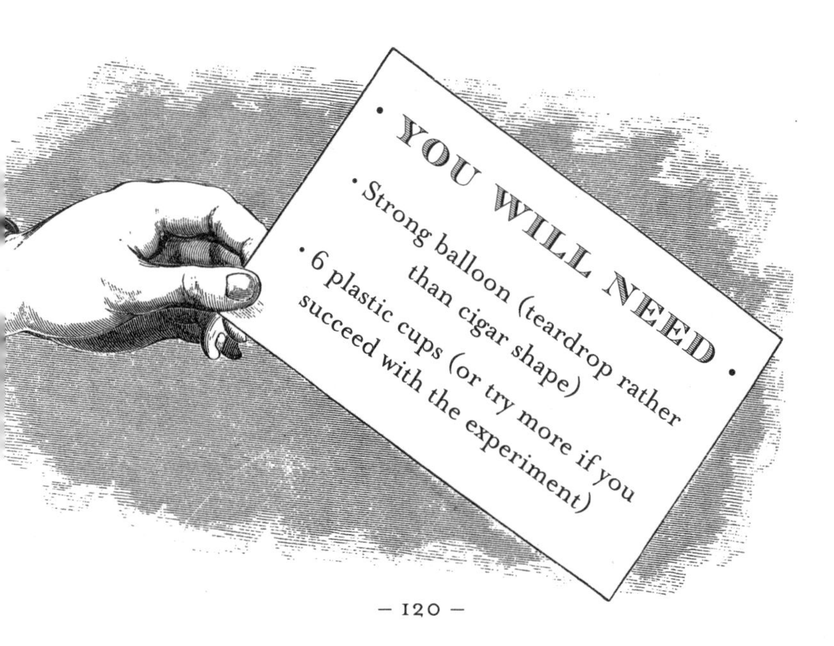

METHOD

1. Blow up the balloon until it's about the size of a grapefruit. Pinch it shut temporarily.

2. Wet the rims of two plastic cups.

3. Press the cups to the sides of the balloon and continue to inflate it. The cups should remain fixed to the balloon.

4. Try repeating the process – pinch, wet, press and inflate – seeing how many balloons you can attach.

THE SCIENTIFIC EXCUSE

As mentioned in the introduction, this experiment is all about air pressure and its counterpart, suction. Wetting the rims of the cups allows them to stick on initially because of water tension. But by inflating the balloon – and reducing the curvature of balloon coming in contact with the cups – the air pressure inside the cups reduces somewhat because the same amount of air now fills more space. The air pressure outside the cups, however, remains the same. This difference of pressure causes the cups to be pushed into the balloon.

TAKE CARE!

Some people obviously have a knack for succeeding with this experiment, but even the least experienced experimenter should be able to get at least two cups to stick. Make sure you keep a little pressure on each cup just after attaching it, so that it will 'take'.

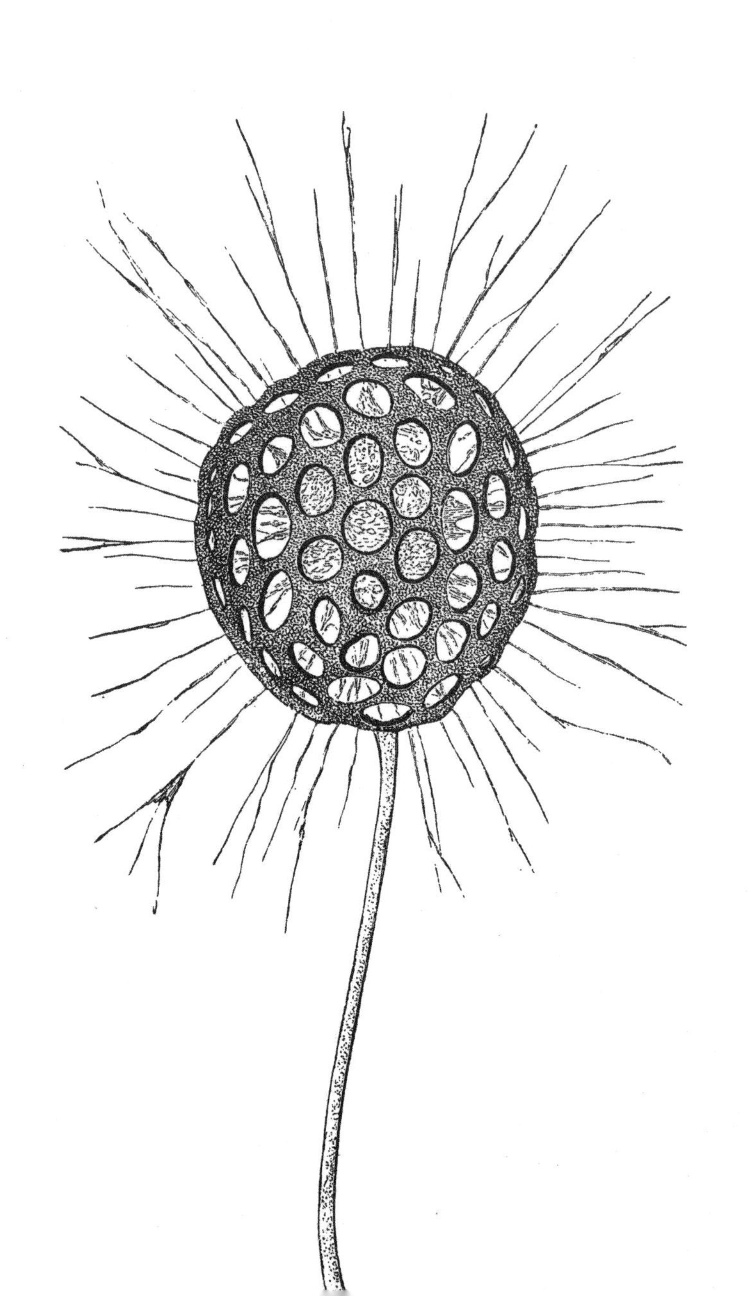

100% NATURAL

Mother Nature is conducting all sorts of experiments – irresponsible or otherwise – all around us, all of the time. The following chapter takes the lid off some of her secrets and leaves the lid firmly shut on those that need darkness and time in order to work out. Flex your scientific muscles a little and then try your hand at directing plants, creating blubber or turning brittle bones to rubber. How can you do all this? Just act 'naturally' and read on.

Pond-life pals

Bringing smelly, slimy water weeds into the freshly cleaned kitchen? Erm, it's 'all in the cause of science': this experiment helps us learn that plants really are the 'lungs' of the world around us.

You will need

- 3–4 shoots of waterweed (e.g. Canadian pondweed, curly waterweed) from a garden pond
- 3 clothes pegs
- 750 ml–1 litre wide-mouthed glass jar
- 7–8 cm test tube
- Small funnel (to fit easily within jar)
- Matches
- Water

Method

1. Fill the jar with fresh water and add the waterweed.

2. Make sure that your waterweed specimens are still alive by checking to see that they emit bubbles.

3. Clip the clothes pegs to the wide rim of the funnel to form a tripod.

4. Lower the funnel-tripod down so that most of the shoots are inside the cone. Make sure the narrow end of the funnel

is higher than the water level.

5. Place the test tube over this end of the funnel.

6. Wait for 2 hours and then remove the test tube carefully, keeping it upside down.

7. Have someone light a match and then blow it out.

8. With the match-head still glowing, place it inside the test tube. It should ignite.

— THE SCIENTIFIC EXCUSE —

The bubbles coming up from the plants (and trapped by the test tube) are oxygen, a by-product of the food-producing process called photosynthesis. After an hour or two, enough oxygen has been collected to provide the fuel for the combustion of the glowing match.

TAKE CARE!

Make sure that you haven't killed your oxygen-producers on the way from the garden pond (see step 2). Be patient so that enough oxygen builds up. Be careful when handling matches but at the same time remember that the match-head still needs to be glowing somewhat if the oxygen is to do its party piece.

MATCH ALERT!

This experiment involves the use of matches and should be conducted only by a responsible adult.

– The –
RUBBER CHICKEN BONE

Politicians sometimes complain – particularly around election time – of spending too much of their lives on the 'rubber chicken circuit'. We all know what they mean: endless meals with boring speakers while they chew and chew and chew their way through tasteless chicken. But perhaps there's a scientific reason for all this, and maybe it's not simply a cliché? Try this experiment and see.

YOU WILL NEED

- Good-sized thigh or drumstick chicken bone, left over after a meal
- Jar big enough to hold the bone with some room to spare (preferably with lid)
- Vinegar

METHOD

1. Clean the bone thoroughly.

2. Rinse it off under running water.

3. Put the bone in the jar and cover with vinegar.

4. Fasten the lid or other covering on the jar.

5. Let the jar sit for about three days.

6. Carefully remove the bone and rinse under running water again.

7. Examine the bone and see how it feels.

— THE SCIENTIFIC EXCUSE —

We can all remember the advice to drink lots of milk and eat dairy products when we were young because they contained calcium to help build our bones. In fact calcium doesn't just build the bones: it keeps them rigid. And it does the same with chicken bones. Vinegar is a mild acid, but still strong enough (given time) to dissolve the calcium (which is a base) in the chicken bone. After a few days, it's done its job and you've got your real-life rubber chicken!

TAKE CARE!

This experiment is safe and you run a risk only if you spill the vinegar–chicken bone mixture. Make sure it's nowhere near carpets or expensive furniture.

The Solid Liquid

(or is it a liquid solid?)

We all know that liquids will turn to solids when the temperature falls enough and that solids (like ice) will melt when things warm up. But how can you explain something that goes from liquid to solid when you tap it, then decides to be a liquid again when you treat it gently? Find out below.

YOU WILL NEED

- 150 g cornflour
- Measuring jug
- 120 ml water
- Large mixing bowl
- Food colouring

METHOD

1. Add the cornflour to the bowl and mix in a few drops of food colouring.

2. Add the water to the bowl – slowly – and mix it with the cornflour using your fingers. Don't necessarily use all of the water: stop when the powder is all wet.

3. Continue mixing with your finger and then tap the surface of the mixture. If the surface feels solid (even though

the mixture just felt like a liquid when you were mixing), then you've got it right.

4. If it's too powdery, add some more water; if it's too liquid, add more cornflour.

5. Try different ways of handling the mixture – rolling it into a ball and then holding it in your palm, banging it with a mixing spoon, flattening it – and see how it behaves.

6. Ask others to decide what its natural state is: solid or liquid?

— THE SCIENTIFIC EXCUSE —

Here's a clue: the answer to that last question is that the mixture is neither a solid nor a liquid, although it has characteristics of both. It's a type of mixture called a colloid. The tiny particles of cornflour remain suspended in the water. Colloids move strangely, displaying a property known as isotropy. This means that the mixture remains like a fluid until agitated. At that point it behaves like a solid.

— TAKE CARE! —

There are no real risks here, apart from those associated with making a mess on the kitchen table.

Self-seal Sandwich Bag

It's great that a sandwich bag can keep your lunch fresh for hours, but isn't it a hassle having to pinch the top, get rid of the excess air and then pull the zip all the way across? Surely there must be some way of saving us all that effort. Could this experiment be a step in that direction? Let's hope so – we might see an end to the condition of chronic 'sandwich-bag tug' fatigue.

You will need:

- One zip-lock sandwich bag
- Three sharp pencils
- Water

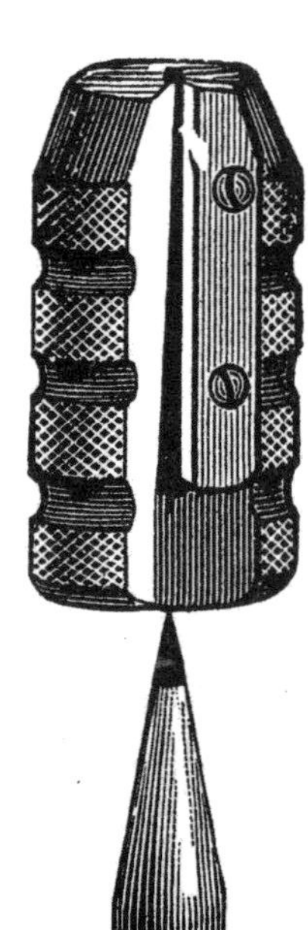

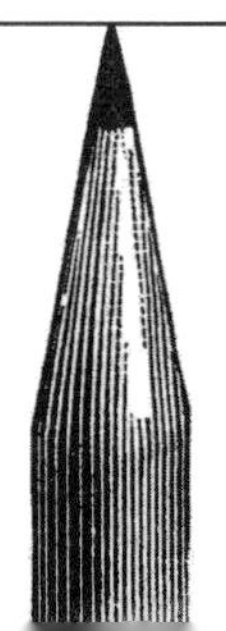

METHOD

1. Fill the sandwich bag about 2/3 of the way with water.

2. Seal the top of the bag as you would normally.

3. Holding the top of the bag, pierce the side with a pencil below the water level and continue until it emerges from the far side of the bag.

4. Repeat step 3 with the second and third pencils.

5. Note how the sandwich bag seals each hole – at least while the pencil is inside!

THE SCIENTIFIC EXCUSE

Rather like the rubber in the unpoppable balloon (p. 112), the plastic in many sandwich bags is naturally elastic – up to a point. Having been punctured by a sharp object, the plastic will form a seal around the edge of the hole. Unfortunately, this power doesn't quite extend to plugging the hole again when the pencil is removed, so be warned.

TAKE CARE!

It's best to do this experiment either outside or by a sink or basin – the pencil removal can get a little messy if you're not careful. If you're feeling ecologically sound, you could be less wasteful by using a sandwich bag that a pupil has brought home from school (having eaten his or her lunch).

Roses are red?

Clever florists in Ireland and in some American cities manage to get green carnations for St Patrick's Day each year. Did you ever wonder how? Here's your chance to be let in on the secret.

YOU WILL NEED

- 2 fountain-pen ink cartridges (red and green)
- 2 test tubes
- 300-ml drinking glass
- Water
- A white-petalled flower (e.g. rose or carnation) with a 15–20-cm stem

METHOD

1. Pour each ink cartridge into a separate test tube.

2. Dilute the ink in each tube by half-filling with water.

3. Carefully slit the stem of the flower so that its base has two halves, each as long as a test tube. Above the slit, the stem should remain whole.

4. Put one stem-half in each test tube and place the tubes (still holding the stem) inside the glass.

5. Keep the tubes upright for several hours.

6. The petals will have changed colour gradually – part becoming red and the other part turning green.

– THE SCIENTIFIC – EXCUSE

The diluted ink is mainly water, which the plant needs for nourishment. It travels through the narrow channels that transport water and nutrients to the different parts of a plant. The colouring (in reality, tiny particles of solid) hits the end of the line when it reaches the petals, although the accompanying water either is used by the plant or evaporates from the surface of the petals. What remains is the distinct colour.

—— TAKE CARE! ——

Make sure the split part of the stem is mainly submerged, since it could allow too much air in otherwise, killing off the plant and ruining your experiment.

– The – Runaway Plant

This wonderful 'stand back and watch' experiment highlights one of the most basic – and at the same time amazing – properties of plants. Just like that seedling that pokes its way through tarmac, the growing potato will work its way through the maze you've built, just to find the nearest source of light. If only we were all so resourceful and persistent!

YOU WILL NEED

- Shoebox
- Sprouting potato
- 60 cm x 60 cm stiff card
- Small flowerpot (just smaller than the height of the shoebox)
- Moist soil
- Scissors

METHOD

1. Measure the inside dimensions (height and width) of the shoebox and cut three shapes of card to that size. These will act as partitions inside the shoebox.

2. Cover the potato with moist soil in the flowerpot.

3. Check that each card shape fits snugly in the shoebox as vertical barriers.

4. Cut a hole in each partition, just wider than the diameter of the potato sprout. Each hole should be in a different position on the card from the others.

5. Cut a fourth hole on one of the upright ends of the shoebox.

6. Slide the three partitions into the shoebox, spaced apart regularly. Put the flowerpot, with its potato, in the shoebox at the bottom end furthest from the end with the hole.

7. Replace the lid carefully on the shoebox, making sure that no light can get into it. Put the shoebox on a surface where light will shine through the top.

8. Leave the box and after a few days you will see the shoot edging through the hole in the top end of the box.

9. Open the lid to see the path that the shoot took to get there.

— THE SCIENTIFIC EXCUSE —

You've just demonstrated a principle called tropism, which explains how a plant reacts to an outside stimulus. Tropism can take different forms: the roots of a plant going down – or out – to find water demonstrate one sort. This experiment demonstrates another, called heliotropism, which describes a plant's efforts to find light.

TAKE CARE!

This is a safe experiment, and holds no concerns for those conducting it.

Sunny Exposure

Just why is it that plants seem to need so much light? Can they get by without it and still look their best? Well, we're not going to tell you. But we've got a way of letting you find out for yourself, provided you're patient enough to wait a week.

YOU WILL NEED

- Healthy, good-sized geranium plant
- Scissors
- Several sheets of black construction paper
- Masking tape

METHOD

1. Cut two squares of construction paper so that each is slightly larger than the size of a geranium leaf.

2. Draw a heart shape in the middle of one of these squares and cut it out carefully, leaving the border of the square intact.

3. Choose a leaf on the plant and carefully sandwich it between the two squares of paper, placing the square with the cut-out heart on the top. Secure the borders of the two squares together with tape.

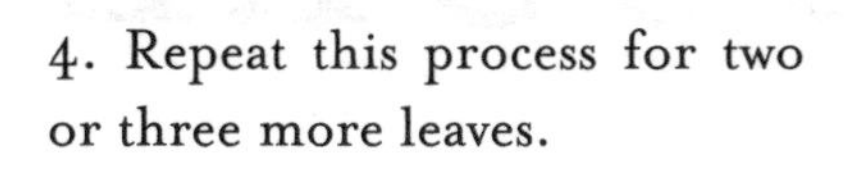

4. Repeat this process for two or three more leaves.

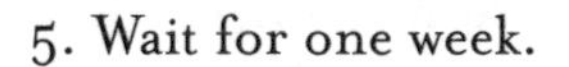

5. Wait for one week.

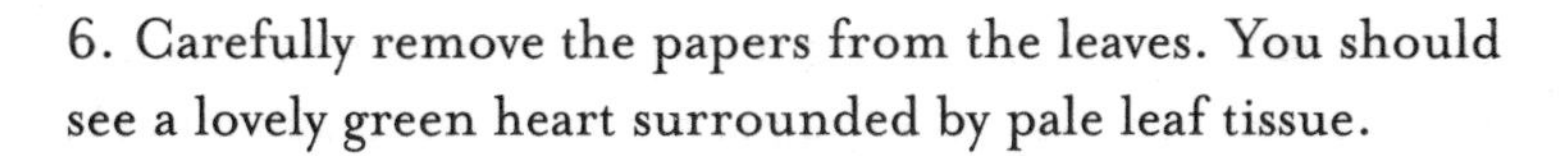

6. Carefully remove the papers from the leaves. You should see a lovely green heart surrounded by pale leaf tissue.

— THE SCIENTIFIC EXCUSE —

The green that we associate with plants (and found in healthy geranium leaves) comes from a chemical called chlorophyll. This chemical is produced in the chemical process known as photosynthesis (the way that plants produce their own food, using light as an ingredient). This experiment has left only part of the leaves exposed to the light – which explains the bright green hearts once you remove the papers. The areas that had been covered had changed to pale green, yellow or – in extreme cases – white.

TAKE CARE!

Be gentle when attaching and removing the paper from the leaves, which are delicate.

DIY BLUBBER

Polar bears, whales and seals can live most of their lives in sub-freezing air or near-freezing waters, yet each of these species is a mammal. And mammals are warm-blooded creatures that need to create and preserve heat in order to survive. Here's a chance to experience the secret of these Arctic survivors. And that secret is called blubber.

YOU WILL NEED

- 250 g vegetable shortening (such as Trex) or lard
- 4 zip-lock sandwich bags
- Masking tape
- Timer (or watch with second hand)
- Cold water and ice cubes
- Sink or deep bucket

METHOD

1. Add the shortening or lard to an open sandwich bag.

2. Turn a second sandwich bag inside out and insert it in the first bag.

3. Zip the inside bag to the outside bag so that the shortening or lard is between them.

4. Tape any gaps where the bags join.

5. Connect the other two sandwich bags together in the same way, only without any shortening or lard. (Each pair of bags becomes a 'mitten'.)

6. Fill the sink or bucket with cold water and add some ice cubes to lower the temperature further.

7. Put both mittens on, start the timer and submerge the mittens in the cold water.

8. Time how long you can keep each mitten – 'blubber' and normal – in the water before you have to pull it out.

— THE SCIENTIFIC EXCUSE —

The key to survival in very cold temperatures is preserving body heat. Blubber, the thick layer of fat under the skin of whales and seals, does that by insulating the body. In effect, it blocks the flow of heat from the body to the outside. In this experiment, the shortening or lard (petroleum jelly would work as well if you had enough of it) insulates your hand in the same way.

TAKE CARE!

Don't try to break any records for holding your 'unblubbered' hand in the water. It should be enough to notice how much the blubber helps keep the other hand warmer.

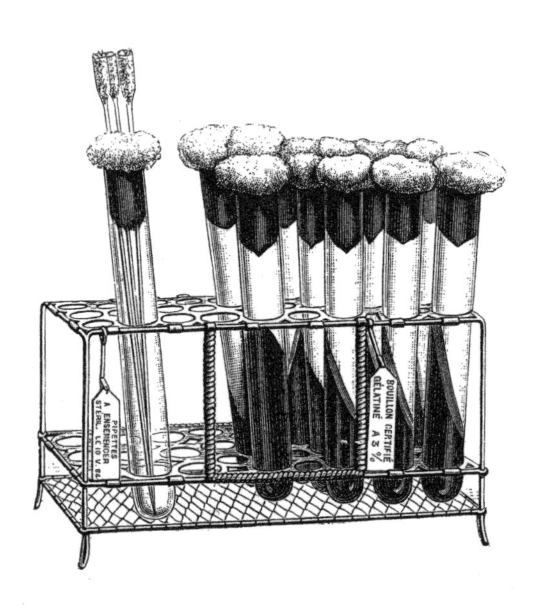
PIPETTES
A ENSEMENCER
STÉRIL. LE 10 V. 84
BOUILLON CERTIFIÉ
GÉLATINÉ A 3 %

Mad Science

The following pages contain some real show-stoppers – experiments that might give you the aura of a Transylvanian scientist with his sidekick Igor. Even with everyday objects such as balloons, paper towels, coins or playing cards, you'll be able to generate gasps from your audience. But don't forget: your bag of tricks comes labelled 'scientific principles'. And the key players behind the scenes are electrical currents, air pressure, chemical reactions and molecular movement.

My cup Overfloweth?

Sometimes people need a demonstration that will show them that the same hard-to-master principles that underpin chemistry can be downright entertaining – or mystifying. See if you can work out why the cup doesn't overflow before you read the scientific excuse. For the best effect, make sure that a member of the 'audience' spoons out the sugar: that way, they won't suspect you of trickery.

YOU WILL NEED

- 300-ml drinking glass
- Piping-hot tap water
- Large jug
- Icing sugar
- Teaspoon

METHOD

1. Set the glass on a table and run the tap until the water is really hot.

2. Fill the jug with hot water, and then pour water from the jug into the glass; slow the flow as you near the top so you

can get it completely full.

3. Very carefully (with the slowest flow possible) continue to add water until the surface bulges over the top but doesn't spill.

4. Now measure 1 tsp sugar and add it to the water from just above the water surface. The water shouldn't overflow despite the addition of the sugar.

5. Repeat this process until eventually the water does overflow.

THE · SCIENTIFIC EXCUSE ·

We sometimes forget that there are spaces between the molecules that make up some of the most familiar substances, such as water. In this experiment, the sugar goes into solution with the water and its molecules slip into these gaps between the water molecules. So, although the glass is 'full' (in fact, more than full because of water tension), it still has room for some sugar.

TAKE CARE!

Make sure your teaspoon doesn't pierce the surface of the water when you add the sugar; it could cause the surface tension to rupture and the water to spill.

- The -
BOTTOMLESS
TIN

You know those peanut-shaped pieces of polystyrene that surround delicate objects that arrive by post – and which seem to get everywhere? You can put these to work in a delightful experiment that cries out for an audience. There's a good scientific explanation as well, so you can feel happy on that score. Just gather up those bits of polystyrene – you'll need more of them than you'd ever imagine!

YOU WILL NEED

- Empty tin (soup or pet-food size), rinsed well
- Polystyrene packing pieces (lots: at least six handfuls)
- Nail-polish remover
- Gloves

METHOD

1. Pour nail-polish remover into the empty tin up to a level of about 2 cm. (If you plan on having an audience, do this before they arrive and make them believe that the tin is empty.)

2. Put the tin on a table next to the large pile of polystyrene pieces.

3. Set aside (or ask an audience member to set aside) the amount of polystyrene pieces that should fill the tin.

4. Drop these pieces into the tin one by one.

5. This pile should soon be used up, so start dropping in pieces from the main pile.

6. Unless you had three full sets of china sent special delivery, and you'd saved every last bit of packing, you will probably use up all of the main pile as well. The volume of the used-up pile could be many times that of the tin.

— THE SCIENTIFIC EXCUSE —

This experiment relies on the chemical interaction between polystyrene and acetone, the active ingredient in nail-polish remover. The 'poly' (Greek for 'many') in the word polystyrene refers to the substance of the structure: it's a long chain of chemical 'sub-units'. This chain, or collection, is called a polymer. The acetone dissolves the linking units of this chain. With these links gone (liquefied), there's very little left of the polystyrene apart from air.

TAKE CARE!

Nail-polish remover is safer than neat acetone (its active ingredient), but remember that it's a solvent and an irritant. Wear gloves when doing this experiment, even if you make them seem like part of a magician's outfit! After performing the experiment, leave the tin outside in a safe place for a day or so. The liquid will solidify and can be disposed of easily.

Matchbox Microphone

Well, if impassioned orators at Speaker's Corner in London can mount soapboxes, why can't you use a much smaller box – a matchbox – to get your message across? This experiment gives you not only the chance to tinker for a good cause, but also a glimpse into the bustling world of those 19th-century inventors.

You Will Need

• Graphite 'leads' (from a mechanical pencil)
• Pin or sharp screw (for piercing the matchbox)
• Base of a small matchbox
• 4.5 V battery
• Old pair of earphones or headphones that you won't need again – ideally, with a good length of cord attached.
• 25 cm connecting wire

Method

1. Pierce two holes, about 1 cm apart, on each of the short sides of the matchbox base. Make sure the holes line up at opposite ends.

2. Break off three lengths of lead. Two of these should be about 1.5 cm longer than the length of the matchbox, and the third should be 2 cm long.

3. Carefully scrape some of the surface off each 'lead'.

4. Insert the two long leads, scraped side up, through the sets of holes in the matchbox and lay the short length (scraped side down) across them.

5. Starting at the end of the earphones cord, separate the two strands for about 15 cm.

6. Connect one of the strands to one of the battery flaps and the second strand to one of the protruding pencil leads.

7. Use the connecting wire to connect the other pencil lead and the second battery flap (thereby completing a series circuit).

8. Have someone hold the earphone and pick up the matchbox carefully – holding it horizontal all the time.

9. Speak into the matchbox – your partner should hear your voice through the earphones.

— THE SCIENTIFIC EXCUSE —

This experiment does call on the basic principle of all microphones. When you speak into the box, it vibrates. This movement is passed on to the leads, which cause the current to flow unevenly. The reverse happens in the earphones: the stop–start in the current causes vibrations, which are translated to sound.

TAKE CARE!

Make sure you can part with the headphones or earphones, since it's much harder to reinstate them than take them apart (and there's not much long-term call for matchbox communications in the modern world).

The NON-LEAK LEAK

Is this an experiment or is it really a trick? Well, it works well as either – or both. Sure, it's a neat demonstration of how a couple of basic physical forces can team up in novel ways. Plus, it's a great way to let the family know-alls get their comeuppance. Anyway, there's an obvious retort to anyone who complains about it: 'Don't be so wet.'

YOU WILL NEED

- Clear plastic 1-litre bottle (screw top)
- Permanent marking pen
- Drawing pin

METHOD

1. Use the permanent marker to write a prominent warning such as 'Do not open' on the bottle, about a third of the way up.

2. Use the drawing pin to poke a series of tiny holes in the bottle at the base of the hand-written letters.

3. Place the bottle in a sink or pan.

4. Fill the bottle and, if possible, keep the water running as

you carefully tighten the screw top.

5. Holding the bottle by the lid, place it carefully on a kitchen counter, with the handwritten warning facing out.

6. Be patient and wait for someone to come along. Plead ignorance if they ask about the warning and watch as they become curious and open it.

7. Water will gush from the series of tiny holes you made with the drawing pin.

— THE SCIENTIFIC EXCUSE —

Everything in this experiment depends on gravity and air pressure. The two need to work in tandem to cause the water to flow out of the bottom of the bottle. When the bottle has been filled, without even the slightest amount of air at the top, gravity is unable to force the water downwards because it needs the air – and air pressure – to act as an intermediary. The same force (air pressure) is holding the water back from flowing through the pin holes. When the cap is removed, gravity combines with air pressure at the top of the bottle to overcome the inward air-pressure force at the holes. The result is a sudden spouting of water on the hapless person who disregarded the warning.

TAKE CARE!

The trick here is to not let the bottle leak too soon. Cover the holes as you fill the bottle right to the top, to ensure that there's no air left. Holding the bottle by the cap as you move it means that it won't get squeezed – another way of producing 'premature ejaculation'.

– The – Shocking Truth

It's a treat when a harmless everyday object – in this case, a jumper – can have the starring role in a demonstration of an eye-popping scientific principle. So, find yourself a cosy sweater and let those sparks fly!

You Will Need

- Woollen jumper (the thicker the better)
- Carpeted flooring

Method

1. Perform this experiment at night or with curtains drawn.
2. Close the door of the room and turn off the light.
3. Slowly rub the jumper along the carpet.
4. Watch as sparks fly!

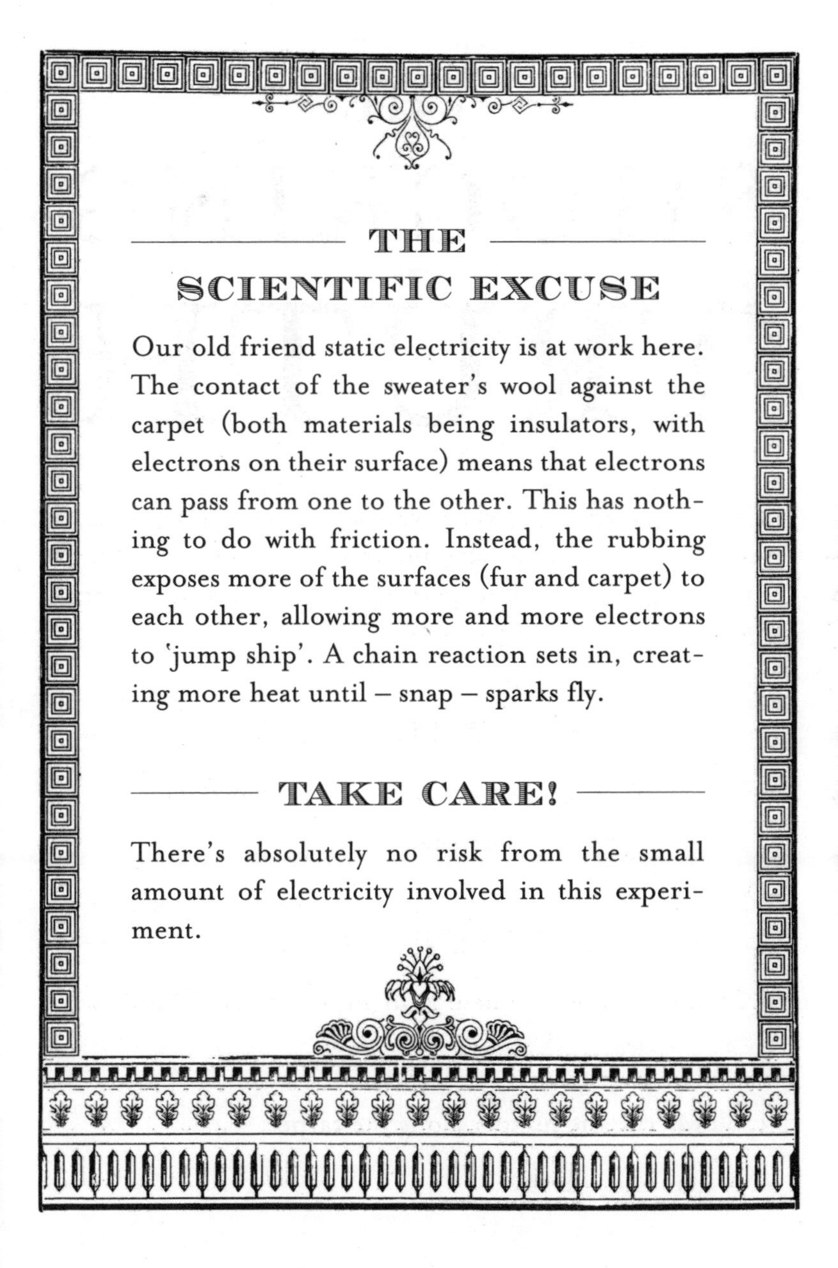

THE SCIENTIFIC EXCUSE

Our old friend static electricity is at work here. The contact of the sweater's wool against the carpet (both materials being insulators, with electrons on their surface) means that electrons can pass from one to the other. This has nothing to do with friction. Instead, the rubbing exposes more of the surfaces (fur and carpet) to each other, allowing more and more electrons to 'jump ship'. A chain reaction sets in, creating more heat until – snap – sparks fly.

TAKE CARE!

There's absolutely no risk from the small amount of electricity involved in this experiment.

– The –
BALLOON BANSHEE

This devilish experiment is so easy that you could start it off – unnoticed – at a children's birthday party. Who could imagine that a colourful party balloon could be so … spooky!

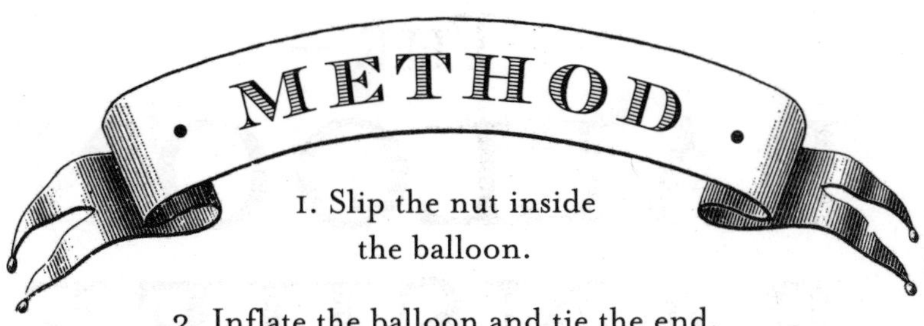

1. Slip the nut inside the balloon.

2. Inflate the balloon and tie the end.

3. Cup one hand over the tied end of the balloon.

4. Tilt the balloon down and begin to move it in a circular path so that the nut inside rolls around.

5. You should start to hear the eeriest, spookiest sound.

THE SCIENTIFIC EXCUSE

The movement of the nut creates vibrations along the balloon's surface. The vibrations cause the air to vibrate both inside and outside the balloon. And vibrating air is another way of describing sound.

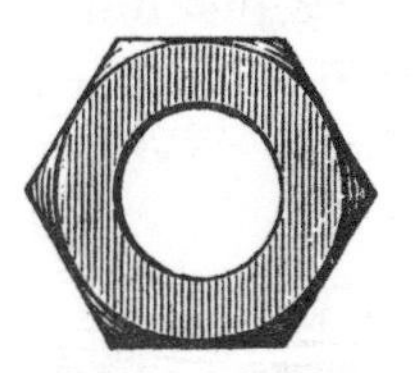

TAKE CARE!

This is a quick, low-risk experiment. Just make sure your nut doesn't have any sharp or rough edges, which could burst the balloon before you hear a thing.

– The –
BUBBLE CHILD

Everyone, young and old, has blown bubbles, but imagine how even the most blasé child would feel actually being engulfed in a bubble. This is a great warm-weather experiment to perform outside. Have a camera ready to record the outcome, since there are bound to be sceptics when you – or your child volunteer – describe the result. The 'secret ingredient' (glycerine) adds colour and strength to the bubbles.

YOU WILL NEED

- A small paddling pool (minimum 1 m diameter)
- Plastic hula hoop
- Wooden footstool
- Swimming goggles
- 500 ml washing-up liquid
- 750 ml water
- 125 ml glycerine (sold at most chemists)
- 10-litre bucket (ideally with lid)

METHOD

1. Mix the washing-up liquid, water and glycerine in the large bucket to make the bubble mixture. (This mixture can keep – and even improves over time – but you must cover the bucket.)

2. Pour the bubble mixture into the paddling pool.

3. Place the hula hoop inside the pool so it's immersed in the solution.

4. Taking care not to puncture the base of the pool, place the stool inside the hula hoop.

5. Ask the child volunteer to stand on the stool – it's always best to choose the most world-weary of the audience, to watch the transformation.

6. Lift the hula hoop up and over the child: a giant bubble will engulf the volunteer.

— THE SCIENTIFIC EXCUSE —

The washing-up liquid (soap) is the prime ingredient for any blown bubble. Each soap molecule has two halves – one hydrophilic (attracted to water) and one hydrophobic (repelled by water). The bubble is actually a 'sandwich': a layer of water molecules squeezed between two layers of soap molecules. The two enemies of bubbles are water tension and evaporation. The interaction with the soap molecules stretches the water molecules apart, weakening the tension. And the other ingredient, glycerine, forms weak hydrogen bonds with the water, slowing or even preventing evaporation.

TAKE CARE!

Make sure the bubble solution is kept from the child's eyes – a pair of goggles might be a good precaution. Also, bear in mind that the quality of bubble varies with type of water ('soft' water is best) and with weather conditions (the best is humid weather).

Overcoming GRAVITY

You can choose the level of irresponsibility for this experiment, depending on your nerve – or your bank balance. It works fine with a glass, a postcard and water: expenditure about £1.10. Or you could try it with Château Margaux if you felt confident enough. No matter which approach you use, it's always a joy to see the faces of youngsters who've never tried out this classic experiment demonstrating air pressure.

YOU WILL NEED

- Drinking glass
- Playing card or postcard
- Water
- Towel or tub (in case of trouble)

METHOD

1. Fill the glass 3/4 full with water.

2. Place the card on the mouth of the glass, making sure that there's no opening.

3. Pressing the card to the rim, turn the glass over.

4. When the glass is upside down, remove your hand from the card.

5. The card remains attached to the glass and no water leaks out.

— THE SCIENTIFIC EXCUSE —

The simple explanation to this experiment is that it all depends on air pressure. The water inside the glass certainly presses down on the card, but what's surprising is the strength of the air pressure working in the opposite direction – greater than the force of gravity, in this case.

• TAKE CARE! •

The best objects to cover the rim of the glass are light but firm – so the playing card and postcard are ideal. Anything heavier scores well with firmness but might be a little too heavy for the air pressure to do its trick. Another word of warning to remember: don't keep the glass and card overturned for too long. If the card becomes soggy, it deforms. That makes it harder for the air pressure to work, so gravity might suddenly win!

The Magic Napkin

Well, this age-old experiment could just as easily have found its way into a magic book. But at its heart is a basic scientific principle – one of the most basic of those proposed by the great Sir Isaac Newton. No points for working out how this one is irresponsible: you'll know in a flash whether it is or it isn't!

YOU WILL NEED

- Plastic cup full of water
- Paper napkin

METHOD

1. Drape the napkin over the edge of a table or kitchen counter.

2. Put the cup on one corner of the napkin about 2–3 cm from the edge of the table.

3. Grasp the overhanging edge of the napkin securely and pull it quickly from under the cup.

4. You should end up with the napkin in your hand and the cup still in place on the table.

— THE SCIENTIFIC EXCUSE —

This experiment demonstrates the principle of inertia, first observed by Sir Isaac Newton as he formulated his famous Laws of Motion. Intertia describes the tendency of objects to stay at rest – in this case to stay on the table rather than be tugged away with the napkin.

TAKE CARE!

You can decide how risky or irresponsible this experiment is by working out where to do it. If the cup does spill, it means that you haven't pulled quickly enough, or with enough force.

Cash or Charge?

Isn't it shocking the way some people are unable to count out change? Here's an experiment that's just as shocking, only this time the shock comes from the coins piling up, not being counted out.

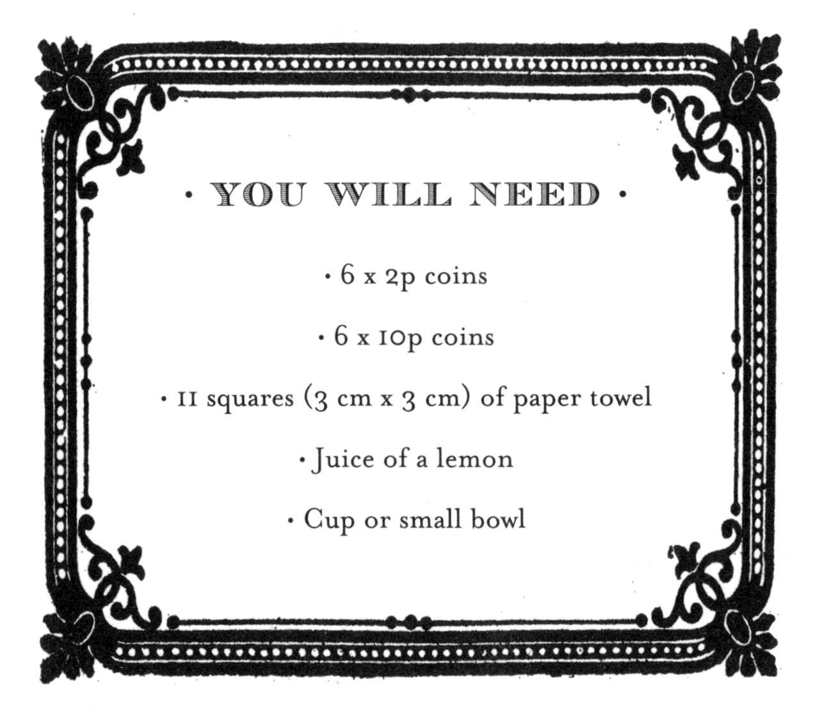

· YOU WILL NEED ·

- 6 x 2p coins
- 6 x 10p coins
- 11 squares (3 cm x 3 cm) of paper towel
- Juice of a lemon
- Cup or small bowl

METHOD

1. Pour the lemon juice into the cup or bowl and add the squares of paper towel.

2. Place the coins in a pile, alternating 2p and 10p coins, and put a lemon-soaked square between each coin. (There should be an exposed coin surface at either end of the pile when you finish.)

3. Moisten the tip of your two index fingers and hold the pile of coins between them.

4. You should feel a small shock.

THE SCIENTIFIC EXCUSE

The shock you felt was a genuine, if slight, electric shock. And electricity, on one level, can be described as the flow of electrons. In this case, the acidic lemon juice releases a positive charge from the copper 2p coins and a negative charge from the 10p coins. That would be the end of it – in other words, the electricity would not 'flow' – if the circuit remained incomplete. By holding the pile of coins between your moistened fingers you have completed the circuit, leading to the electric tingle. Normal batteries, such as those in a car or a torch, operate on the same principle as this 'wet cell'.

TAKE CARE!

The amount of electricity produced here is very slight, so there's no risk involved. Just to make sure it all works, however, make sure the paper towel squares are wet but not dripping. Also, clean coins work better than those that are soiled.

The Non-Drip Document

'Less is more', as the great modernist architect Mies van der Rohe would say, and this simple experiment certainly lives up to the adage. It takes just over a minute to perform, but it gives a lovely demonstration of some basic science. And what's more, the main ingredient – the piece of paper – can become the star attraction. Try using something that looks really important, something that looks as though it would be ruined if even a drop of water got onto it. That should make the wait in the last stage the longest minute your audience will have endured.

You · Will · Need

- Clear drinking glass
- Water
- Sink or large 3–4-litre pot
- A4-sized piece of paper (could be newspaper or, ideally, something that looks important or 'business-like')

METHOD

1. Fill the sink or pot nearly full of water.

2. Loosely crumple the piece of paper.

3. Shove the crumpled paper into the bottom of the empty glass, tight enough so that it won't fall out when the glass is overturned.

4. Plunge the glass upside down into the water, deep enough that the paper appears to be under water.

5. Keep the glass in that position for a minute.

6. Take the glass out of the water and retrieve the paper; dramatically straighten it and hand it round. It will be dry.

THE SCIENTIFIC EXCUSE

The air inside the glass has protected the paper from the water by acting as a barrier. This air is lighter than the water, so it can't flow down through the rim of the glass. At the same time, the water can't flow into the glass because the glass is already full – of air!

TAKE CARE!

This experiment is a revelation even if you use a discarded page from a newspaper that would otherwise be recycled. If you do raise the stakes, make sure you use a piece of paper that seems to be valuable (a photocopy of an invitation or school letter, for example). You don't want to run the risk of having the real thing fall out because you hadn't shoved it in hard enough.

Introduction to Alchemy

All right, we can't pretend to teach you how to find the Philosopher's Stone, but this experiment does touch on one of the main aims of those medieval alchemists – changing the nature of metals. Of course, they were looking for a way of turning base metals such as lead into gold, but here you'll have a chance to turn iron into copper. Or at least that's what you'll seem to have done.

— You Will Need —

- 12 x 1p or 2p coins (dull ones work better)
- Spoon
- One ungalvanised iron nail
- Ceramic or plastic bowl
- 150 ml vinegar
- 1 tsp salt
- Paper towel

Method

1. Add the vinegar to the bowl and stir in the salt.

2. Place the dull coins in the bowl (adding a little more vinegar if they aren't fully submerged).

3. Leave the coins in the bowl for 5 minutes.

4. Retrieve the coins from the bowl using the spoon and let them dry on the paper towel (do not dump out the vinegar/salt mixture).

5. Add the nail to the bowl of liquid and keep it there for 30 minutes. Observe the tiny bubbles forming around it.

6. Carefully take the nail out of the bowl and set it on paper towel to dry. It should appear to be copper!

THE SCIENTIFIC EXCUSE

In the first stage of the experiment, the vinegar solution 'cleaned' some of the copper off the coins. This copper remained in solution (with the vinegar and salt), which in turn had a chemical reaction with the iron on the surface of the nail. As part of this reaction, a chemical exchange left a copper coating on the nail. This type of chemical addition of a metallic layer is called plating; you have created a copper-plated nail. Next stop gold!

TAKE CARE!

This experiment works best if the concentration of copper (from the coins) is relatively high – so make sure you don't overdo the amount of vinegar you use. Don't choose too big a bowl, since the amount of liquid needed to cover the coins would weaken the solution.

A good head for lights

The pun in this title might make you groan, but the nifty experiment below has all the hallmarks of a classic demonstration: it's quick, easy and scores high on the 'ooooh' scale. And what's more, there's some really interesting science at its heart. For best results, perform the experiment after dark, so the effect will appear that much spookier.

YOU WILL NEED

- Balloon
- Fluorescent (strip) light bulb
- Volunteer with a good head of hair
- Second volunteer to hold the bulb

METHOD

1. Inflate the balloon and tie it shut.

2. Have one volunteer hold the bulb carefully, with the metal contact pointing outwards.

3. The person with the balloon should rub it vigorously against his or her head for 10 seconds and then touch the metal contact of the fluorescent bulb.

4. The bulb should give off a ghostly glow.

— THE SCIENTIFIC EXCUSE —

The gas inside a fluorescent light bulb becomes 'excited' when even a small amount of electricity flows through it. Rubbing the balloon against hair creates static electricity, so when the balloon touches the metal contact, electrons flow into the bulb. This modest electrical charge excites the low-pressure gas inside, which in turn excites the phosphorous coating of the inside of the bulb. That last stage produces visible white light.

TAKE CARE!

Make sure that the volunteer handling the fluorescent bulb is careful and responsible.

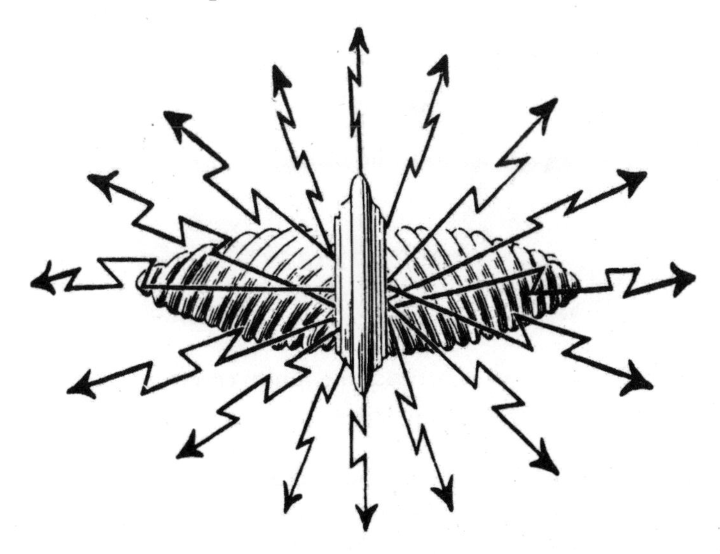

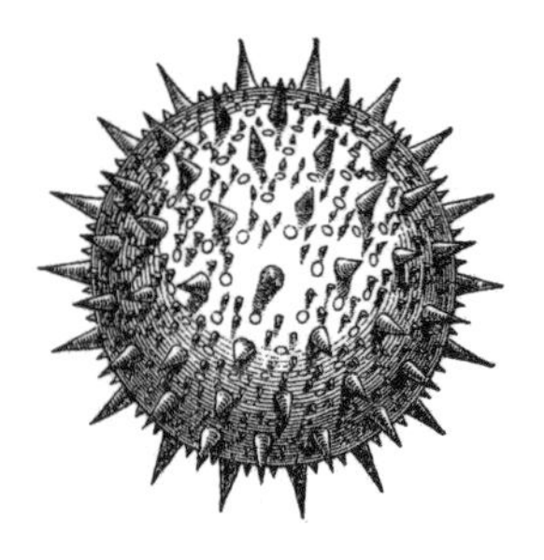

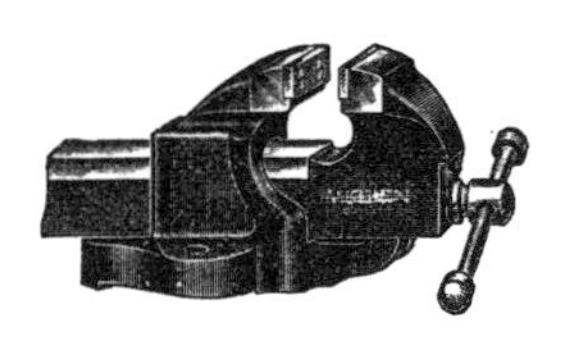

Handy Oil Can
THREE IN ONE OIL
3 IN 1
PREVENTS RUST
LUBRICATES
CLEANS AND
POLISHES
TALKING MACHINES
SEWING MACHINES
TYPEWRITERS &
ELECTRIC FANS
RAZORS & STROPS
FIRE ARMS
MAGNETOS, COMMUTATORS
CASH REGISTERS & LAWN
MOWERS LIGHT MACHINERY, ETC
PIANOS, FURNITURE & WOODWORK
Manufactured by
THREE IN ONE OIL COMPANY
NEW YORK U.S.A.
PRICE 25 CENTS

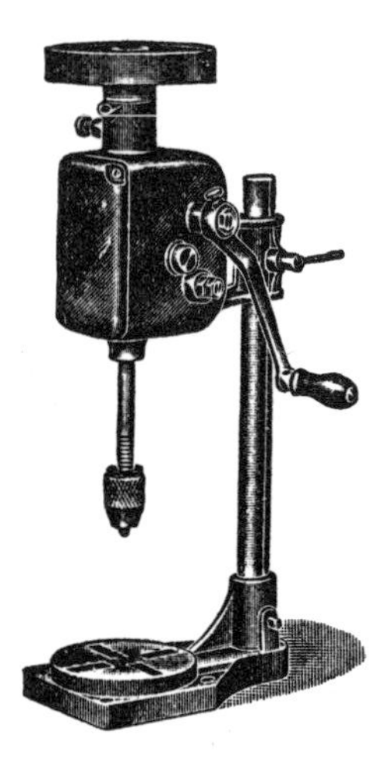

AT A GLANCE

Some of the 65 experiments in this book can be done almost in an instant; others take up to several days to achieve their dramatic effect. The following list groups them in order of time taken, starting with those that can be done most quickly.

FLASH IN THE PAN

less than 2 minutes

5-MINUTE WONDERS

2–5 minutes

On the hour

up to 1 hour

The 8-hour day

1–8 hours

Going the distance

a full day or more

Acknowledgements

I would like to thank the following individuals for their support and inspiration at every stage of this project. Their contributions – including direct participation, suggestions, reactions and scientific explanations – added depth and breadth to the book:

Frank Ciccotti, Gregory Etter, Peter French, Dr Gary Hoffman, Benjamin Joyce, Dr Peter Lydon, William Matthiesen, Ian McChesney, Dr Sarah Morse, Peter Rielly, Susan Roeper, Elizabeth Stell

In addition, I owe a debt of gratitude to the following companies and organisations:

Berkshire (Massachusetts) Film and Video, The Corsham Bookshop, Energy for Sustainable Development Ltd, ESD Ventures

GOOGLIES, NUTMEGS & BOGEYS

The Origins of Peculiar Sporting Lingo

BOB WILSON

Googlies, Nutmegs & Bogeys

The Origins of Peculiar Sporting Lingo

BOB WILSON

Have you ever flashed at a zooter in the corridor of uncertainty while on a sticky dog? Maybe you've seen someone hit a mulligan out of the screws to grab a birdie at Amen Corner?

The world of sport has its own language, wonderfully rich in strange words and phrases, whose origins often stretch back centuries. Veteran BBC presenter and football legend Bob Wilson has written this brilliant illustrated guide to the fascinating true meanings, heritage and evolution of the great sporting terms we use today.

'A lesson in the language of sport from a man that should know'

Kevin Keegan OBE

'The daddy of male loo-reading books ... Bob Wilson's compendium should be sent out by right to every lad when he reaches the age of assent, in exactly the same way centurions receive a telegraph from the Queen.'

Scotland on Sunday

'Wilson conducts an engaging romp through sport's more colourful terminology ... buy it to be entertained.'

Independent

BOB WILSON is the former presenter of nine football World Cups, as well as BBC *Grandstand*, *Football Focus*, *Sportsnight* and *Match of the Day*. He also presented the most watched football match in British TV history, with an audience of over 26 million. Bob originally qualified as a teacher before being lured away to play football for Scotland and Arsenal, with whom he won the European Fairs Cup and the League and FA Cup double.

UK £9.99

ISBN13: 978-1840467-74-1

THE GRAND PRIX COMPANION

ALAN HENRY

Alan Henry has been part of the F1 paddock for 34 years, covering 536 races and almost 2 million miles in the process. Now he has compiled *The Grand Prix Companion* – a true insight into the world of Formula 1 from a man who should know.

Although principally a compendium of the weird and wonderful facts thrown up by a statistic-obsessed sport, it's also laced with the anecdotes and memoirs of one of the pit-lane's longest-serving and most respected insiders. Among the nuts, bolts and wind tunnels, discover a thing or two about the legendary negotiating methods of Bernie Ecclestone, and the peculiar antics of the stars of the sport past and present, including the true extent of the infamous nocturnal exploits of Alan's friend, the late and great James Hunt.

In short, essential reading for any Formula 1 fan.

'I've known Alan Henry since before I moved up into F1 and he probably has a bigger fund of racing anecdotes than anybody in this business. I am sure anybody who has even a passing interest in this spectacular sport will find something to absorb them in this book.'

Niki Lauda, three-times Formula 1 World Champion

'A unique insight into the sport ... a must read for anybody whose juices flow at the prospect of F1.' *Sunday Times* (Book of the Week)

ALAN HENRY covered his first Formula 1 race way back in 1973. He is now editor of *Autocourse*, editor-at-large for *F1 Racing* and Grand Prix correspondent for *Autocar*, and has been motor racing correspondent for the *Guardian* since 1987. He is one of the most respected motorsport journalists in the world.

UK £9.99

ISBN13: 978-1840467-96-3

1